Diet, Nutrition, and Cancer:
A Critical Evaluation

Volume II
Micronutrients, Nonnutritive Dietary Factors, and Cancer

Editors

Bandaru S. Reddy, Ph.D.
Member and Associate Chief, Division of
Nutrition and Endocrinology
Head, Nutritional Biochemistry Section
Naylor Dana Institute for Disease Prevention
American Health Foundation
Valhalla, New York

Leonard A. Cohen, Ph. D.
Head, Nutritional Endocrinology Section
Division of Nutrition and Endocrinology
Naylor Dana Institute for Disease Prevention
American Health Foundation
Valhalla, New York

CRC Press
Taylor & Francis Group
Boca Raton London New York

CRC Press is an imprint of the
Taylor & Francis Group, an **informa** business

First published 1986 by CRC Press
Taylor & Francis Group
6000 Broken Sound Parkway NW, Suite 300
Boca Raton, FL 33487-2742

Reissued 2018 by CRC Press

Library of Congress Cataloging in Publication Data
Main entry under title:

Diet, nutrition, and cancer.

   Includes bibliographies and indexes.
   Contents: v. 1. Macronutrients and cancer --
v. 2. Micronutrients, nonnutritive dietary factors,
and cancer.
   Cancer--Nutritional aspects. I. Reddy,
Bandaru S. II. Cohen, Leonard A. [DNLM: 1. Diet--
adverse effects. 2. Neoplasms--etiology. 3. Neoplasms
--prevention & control. 4. Nutrition, QZ 202 D5653]
RC268.45.D54   1986   616.99'4   85-15172
ISBN 0-8493-6332-2 (v. 1)
ISBN 0-8493-6333-0 (v. 2)

A Library of Congress record exists under LC control number: 85015172

ISBN 13: 978-1-315-89228-3 (hbk)
ISBN 13: 978-1-351-07138-3 (ebk)

Visit the Taylor & Francis Web site at http://www.taylorandfrancis.com and the
CRC Press Web site at http://www.crcpress.com

PREFACE

Although the concept that diet and nutrition might influence cancer is not new, until recently this relationship has received surprisingly little detailed attention. During the 1930s, a number of laboratories, including that of Tannenbaum, were interested in the possible influence exerted by nutritional factors on susceptibility to cancer, but the question soon lost the interest of both the scientific and lay community.

During the past 2 decades, renewed interest in nutritional carcinogenesis has developed. Epidemiologic studies have investigated the incidence pattern between and among population groups, differences in the rates of the disease between the sexes, changes in disease rates over time, demographic and socioeconomic distribution of diseases, effects of migration, and the dietary habits of different population groups, and have led to the conclusion that nutritional factors play a significant role in the etiology of certain types of cancer. However, it must be recognized that the correlation between nutritional factors and certain forms of cancer does not prove causation. Many factors may be necessary for cancer causation, but the modification of only one of the contributing factors, such as diet, may be sufficient to retard the chain of causative events.

Studies in experimental animal models also point to dietary factors as important modulators of certain types of cancer. These studies have generally shown that increased macronutrient intake, especially fat, and certain micronutrient deficiencies lead to increased in tumor incidence in several organ sites, whereas diet restriction and dietary excess of certain micronutrients lead to a lower tumor incidence.

These 2 volumes bring together a wide variety of studies concerning the role nutrition plays in the etiology of various types of cancer, namely, cancer of the esophagus, upper alimentary tract, pancreas, liver, colon, breast, and prostate. The purpose of each chapter is to provide a critical interpretive review of the area, to identify gaps and inconsistencies in present knowledge, and to suggest new areas for future reasearch. Scientifically valid data supporting an association between nutrition and cancer comes from three sources: epidemiology, clinical studies, and experimental studies in laboratory animal models. Throughout the volumes, attention is given to the potential and limitations of each discipline; and the need for closer cooperation between epidemiologists, clinicians, and experimentalists is emphasized. Specific areas of concern include extrapolation of data from animal models to humans, methods of diet evaluation, the formation and occurrence of mutagens in cooked food, and the role of naturally-occurring inhibitors of carcinogenesis.

We have tried to present in 19 chapters (9 chapters in Volume I, 10 chapters in Volume II) a comprehensive view of nutrition's role in cancer. The broad coverage of diet, nutrition, and cancer provided by these chapters is intended to serve both as an introduction to readers unfamiliar with the field, as well as a source of new information for researchers. It is indeed hoped that these 2 volumes will promote a better understanding of the role of nutritional factors in the induction and inhibition of cancer, and that this understanding will lead to a reduction in cancer rates in the current generation and the prevention of cancer in future generations.

Obviously, the compilation of these volumes could not have come about without the cooperation of the various authors. We most sincerely thank each of the authors for their contribution and continued assistance in submitting and editing the manuscript.

THE EDITORS

Bandaru S. Reddy, D.V.M., Ph.D., is presently a Member and Associate Chief of the Division of Nutrition and Endocrinology, Naylor Dana Institute for Disease Prevention, Valhalla, New York, and Research Professor of Microbiology at New York Medical College, Valhalla, New York. He received a degree in Veterinary Medicine in 1955 from the University of Madras, India, an M.S. in 1960 from the University of New Hampshire, and Ph.D. in 1963 from Michigan State University. He then spent 8 years as a faculty member at the Lobund Laboratory of the University of Notre Dame, Notre Dame, Indiana.

His current research interest is diet, nutrition and cancer, with particular emphasis on large bowel cancer. This is an important area of investigative research that may contribute to primary and secondary prevention of cancer. He has published about 160 papers on this and related subjects.

Dr. Reddy is a member of the American Institute of Nutrition, American Association of Pathologists, American Association of Cancer Research, Society for Experimental Biology and Medicine, Association of Gnotobiotics, and the Society of Toxicology. During 1979, he served as the President of the Association of Gnotobiotics.

Leonard A. Cohen is presently Head, Section of Nutritional Endocrinology, Naylor Dana Institute for Disease Prevention, Valhalla, New York. He received a Ph.D. from the City University of New York in 1972 in cell biology. Dr. Cohen joined the NDI soon after receiving his doctorate and has dedicated his efforts since then to the elucidation of the mechanism of mammary tumor promotion by dietary fat.

Dr. Cohen is a member of the American Association for Cancer Research, the International Association for Breast Cancer Research, The International Association for Vitamins and Nutritional Oncology, The Tissue Culture Association, and the American Association for the Advancement of Science.

CONTRIBUTORS

Kulbir Bakshi
Staff Officer
Board on Toxicology and Environmental
  Health Hazards
National Academy of Sciences
Washington, D.C.

William S. Barnes
Assistant Professor
Department of Biology
Clarion University of Pennsylvania
Clarion, Pennsylvania

T. Colin Campbell
Professor of Nutritional Biochemistry
Division of Nutritional Sciences
Cornell University
Ithaca, New York

Chandramohan B. Chawan
Assistant Professor
Food Science and Animal Industries
Alabama A and M University
Normal, Alabama

Malford E. Cullum
Research Associate
Department of Food Science and Human
  Nutrition
Michigan State University
East Lansing, Michigan

Richard Czerniak
Associate
Division of Molecular Biology and
  Pharmacology
American Health Foundation
Naylor Dana Institute for Disease
  Prevention
Valhalla, New York

G. David McCoy
Assistant Professor
Department of Environmental Health
  Sciences
Case Western Reserve University
School of Medicine
Cleveland, Ohio

Daniel Medina
Professor
Department of Cell Biology
Baylor College of Medicine
Houston, Texas

Curtis Mettlin
Director
Department of Cancer Control and
  Epidemiology
Roswell Park Memorial Institute
Buffalo, New York

Nancy G. Misslbeck
Lecturer
Division of Nutritional Sciences
Cornell University
Ithaca, New York

Sushma Palmer
Executive Director
Food and Nutrition Board
National Academy of Sciences
Washington, D.C.
Adjunct Assistant Professor
Department of Pediatrics
Georgetown University School of
  Medicine
Washington, D.C.

Sudhanand Reddy Pulsani
Instructor
Department of Food Science and Animal
  Industries
Alabama A and M University
Normal, Alabama

Damanna Ramkishan Rao
Professor
Department of Food Science and Animal
  Industries
Alabama A and M University
Normal, Alabama

Nrisinha P. Sen
Research Scientist
Health and Welfare Canada
Health Protection Branch
Ottawa, Ontario, Canada

Raymond J. Shamberger
Head
Section of Enzymology
Department of Biochemistry
The Cleveland Clinic Foundation
Cleveland, Ohio

John H. Weisburger
Vice President for Research and Director
Naylor Dana Institute for Disease
 Prevention
American Health Foundation
Valhalla, New York

Clifford W. Welsch
Professor
Department of Anatomy
Michigan State University
East Lansing, Michigan

Maija H. Zile
Associate Professor
Department of Food Science and Human
 Nutrition
Michigan State University
East Lansing, Michigan

TABLE OF CONTENTS

Volume I

Volume II

TABLE OF CONTENTS

Chapter 1

RETINOIDS AND MAMMARY GLAND TUMORIGENESIS: A CRITIQUE

Clifford W. Welsch, Maija H. Zile, and Malford E. Cullum

TABLE OF CONTENTS

I. INTRODUCTION

In 1909 Stepp[1] first described a fat-soluble material which proved essential for life. Stepp carried out experiments on extracted "fat-free" diets and was able to show that the extracted food did not allow laboratory mice to survive unless an extract from egg yolk was added to the diet. Additional animal sources of the fat-soluble material were subsequently found in butter-fat and cod liver oil. In 1920, Drummond[2] suggested that the fat-soluble material should be called vitamin A. In 1931, Karrer et al.[3] obtained vitamin A preparations of high purity and activity from fish liver oils and proposed the correct structural formula for vitamin A. By 1942, Baxter and Robeson[4] succeeded in crystalizing pure vitamin A and several of its esters. The synthesis of crystalline vitamin A was first described by Isler et al.[5] in 1947.

The term vitamin A is now used when reference is made to the biological activity of a number of vitamin A active substances. The three most common vitamin A active substances are retinol (vitamin A alcohol), retinal (vitamin A aldehyde), and retinoic acid (vitamin A acid). All three compounds contain as common structural units a trimethylcyclohexenyl group and an all-*trans* polyene chain with four double bonds. Collectively, natural vitamin A compounds are important in general growth, growth and differentiation of epithelial tissues, visual function, and reproduction.

A relationship between vitamin A and cancer was suggested by Wolback and Howe,[6] who in 1925 first described vitamin A deficiency in detail. In this early study, they also pointed out the similarity between vitamin A deficiency-induced metaplastic lesions in the epithelium of GI, respiratory, and urogenital tracts. In certain organs, these changes are regarded as a first step in neoplastic transformation. One year later, Fujimaki[7] reported development of carcinomas in the stomach of laboratory rats which were fed a vitamin A-deficient diet. These early observations have led to testing of the concept that vitamin A compounds could be used in the prevention of tumorigenesis. In the late 1960s and early 70s a number of laboratories showed that induction of benign and malignant epithelial tumors in laboratory animals could be retarded or prevented by treatment of animals with relatively high doses of a variety of retinoids. To date, the chemopreventive effect by pharmacological doses of retinoids has been demonstrated in a variety of experimental tumor models and organ sites. Carcinogen-induced tumorigenesis in the bladder,[8] cervix,[9] vagina,[9] colon,[10] integument,[11] forestomach,[9] tracheo-bronchi,[12] pancreas,[13] liver,[14] and mammary glands[15] have been suppressed by feeding high levels of the vitamin.

The purpose of this communication is to review the relationship between retinoids and mammary gland tumorigenesis in rats, mice, and humans and to examine mechanism(s) by which retinoids might influence this neoplastic process. For those interested in the relationship between dietary vitamin A and tumorigenesis at organ sites other than the mammae, a number of reviews are available.[16-21]

II. RETINOIDS AND MAMMARY GLAND TUMORIGENESIS

A. Rats

The first direct evidence of a significant inhibitory effect of high levels of dietary retinoids on the development of mammary tumors in experimental animals was reported by Moon et al.[15] in 1976. Female Sprague-Dawley rats were intubated with a single dose of 7,12-dimethylbenzanthracene (DMBA) and 7 days later fed a standard laboratory chow containing retinyl acetate. Daily consumption of retinyl acetate was adjusted to either 1.0 or 2.5 mg/rat. The incidence of carcinomatous and benign mammary tumors 7 months after carcinogen treatment was significantly reduced in animals fed the natural retinoid. This study has been repeatedly confirmed in many laboratories using a variety of mammary gland carcinogens, i.e., DMBA, *N*-methyl-*N*-nitrosourea (MNU) and benzopyrene (BP), and dietary retinoids,

Table 1
INHIBITION OF CHEMICAL
CARCINOGENESIS OF THE RAT
MAMMARY GLAND BY DIETARY
RETINOIDS[a]

Carcinogen	Retinoid	Ref.
DMBA	Retinyl acetate	15
	Retinyl acetate	22
	Retinyl acetate	23
	Retinyl acetate	24
	Retinyl acetate	25
	Retinyl methyl ether	26
	4-HPR	27
MNU	Retinyl acetate	28
	Retinyl acetate	29
	Retinyl acetate	30
	Retinyl acetate	31
	Retinyl methyl ether	32
	4-HPR	33
	4-HPR	27
	13-*cis*-retinoic acid[b]	32
BP	Retinyl acetate	34

[a] Female Sprague-Dawley rats were used in all of these studies with the exception of Reference 22 and 34 in which female Lewis rats were used.
[b] Marginally active.

i.e., retinyl acetate, retinyl methyl ether, 13-*cis*-retinoic acid and *N*-(4-hydroxyphenyl)retinamide (4-HPR) (Table 1).[22-34]

McCormick et al.[22] initiated a series of studies to determine whether mammary tumorigenesis could be suppressed in DMBA-treated female rats when retinyl acetate feeding was limited to specific time periods during this neoplastic process. Feeding retinyl acetate at -2 to $+1$ weeks, $+1$ to $+30$ weeks, $+1$ to $+12$ weeks, $+12$ to $+30$ weeks, and -2 to $+30$ weeks, where time 0 was the day of DMBA administration, resulted in a striking difference in tumorigenesis between these groups; 30 weeks after DMBA treatment mammary carcinoma incidence was significantly reduced in all groups but one ($+1$ to $+12$ weeks) when compared with placebo-fed controls. The greatest decrease in mammary tumor incidence was seen in the longest treatment group (-2 to $+30$ weeks) but a nearly equal reduction was seen in the group receiving a short retinyl acetate exposure (-2 to $+1$ weeks). In the $+1$ to $+12$ week group, the inhibition of tumor development was temporary; tumor incidence returned to that observed in placebo-fed control rats by 30 weeks postcarcinogen treatment. Nearly identical results were obtained in a subsequent study using a different carcinogen (BP).[34] However, when retinyl acetate feeding is delayed until 6 months after carcinogen treatment, the effectiveness of the retinoid may be reduced or absent.[35]

In all of the cited studies in Table 1, large amounts of dietary retinoids (0.6 to 2.0 mM) were used for successful chemoprevention of rat mammary gland carcinogenesis. Dietary retinyl acetate levels as low as 0.2 mM is not, however, effective in the chemoprevention of rat mammary gland carcinogenesis.[25] This level of retinyl acetate (0.2 mM) still provides $\approx$20 times the daily requirement (IU) of vitamin A! Thus a narrow and continuous high level of the retinoid is prerequisite for successful chemoprevention of this neoplastic process. Retinyl acetate accumulates excessively in the liver, causing mild hepatic toxicity.[33] Whether or not retinyl acetate-induced hepatic toxicity affects this neoplastic process remains to be

Table 2
ENHANCEMENT OF THE INHIBITORY ACTION
OF DIETARY RETINOIDS ON RAT MAMMARY
GLAND CARCINOGENESIS BY OTHER
BIOLOGICAL RESPONSE MODIFERS

Carcinogen	Retinoid	Biological response modifier	Ref.
MNU	Retinyl acetate	Hormone antagonism[a]	29
MNU	Retinyl acetate	Hormone antagonism[c]	31
MNU	4-HPR	Hormone antagonism[c]	27
MNU	Retinyl acetate	Selenium	30
MNU	Retinyl acetate	Hormone antagonism[b]	36
MNU	Retinyl acetate	Immune stimulation	36
DMBA	Retinyl acetate	Selenium	23
DMBA	Retinyl acetate	Hormone antagonism[c]	24
DMBA	Retinyl acetate	Hormone antagonism[b]	25
DMBA	Retinyl acetate	Immune stimulation	25

[a]   Hormone antagonism — CB-154 treatment.
[b]   Hormone antagonism — CB-154 + tamoxifen treatment.
[c]   Hormone antagonism — ovariectomy.

determined. Despite the mild hepatic toxicity, rats fed dietary levels of retinyl acetate up to 1.0 mM have normal body weight gains, normal estrous cycles, and in general appear as healthy as placebo-fed control animals. Certain synthetic retinoids (e.g., 4-HPR) do not accumulate excessively in the liver; therefore little or no hepatic toxicity is observed in these animals.[33] Thus, retinoid-induced prophylaxis of rat mammary gland carcinogenesis can be observed in animals without concurrent hepatic toxicity.

Recent studies have demonstrated that the chemopreventive activities of retinoids in rat mammary gland carcinogenesis can be significantly enhanced by other biological response modifiers, i.e., hormone antagonism, immune stimulation, and the administration of pharmacological levels of selenium (Table 2). In comparing the effectiveness of retinoid feeding, selenium treatment, and hormone antagonism in the prophylaxes of rat mammary gland carcinogenesis, it is clear that hormone antagonism is superior to retinoid feeding.[25,36] Retinoid feeding, on the other hand, appears to be slightly superior to selenium treatment.[28,30] Clearly, the effectiveness of retinoid feeding can be enhanced by either selenium treatment or hormone antagonism. Immune stimulation (cell particulate of mammary carcinoma plus Freund's complete adjuvant), although not effective in suppressing rat mammary gland carcinogenesis when given alone, did significantly reduce mammary carcinoma incidence in rats fed retinyl acetate.[25] A lack of synergism between other immune stimulants (i.e., methanol-extracted residue of *Bacillus Calmette-Guerin*, cell wall skeleton of *Nocardia rubra*, and maleic anhydride-divinyl ether copolymer — MVE-2) and retinoid feeding in the chemoprevention of rat mammary gland carcinogenesis has been reported.[25,37] The comparative efficacies of retinoid feeding, hormone antagonism (inhibition), and immune stimulation and synergism between these biological response modifiers in the chemoprevention of rat mammary gland carcinogenesis was recently reported by Welsch and DeHoog.[25] The results of this study are summarized in Table 3. At termination of study (20 weeks after carcinogen treatment), no mammary carcinomas were observed in rats treated with the combination of retinoid feeding, hormone antagonism, and immune stimulation. In a subsequent study, by 1 year after carcinogen treatment, only 2 mammary carcinomas were observed in a group of 40 rats treated similarly.[36]

Although the effects of dietary retinoids have been extensively examined in carcinogen-

Table 3
RETINOID FEEDING, HORMONE INHIBITION, AND/OR
IMMUNE STIMULATION AND THE PROGRESSION OF
DMBA INDUCED RAT MAMMARY CARCINOMAS

Treatment[a]	Number of rats	Total number of mammary carcinomas 20 weeks after DMBA treatment
Controls	40	81
Immune stimulation (IS)[b]	40	77
Hormone antagonism (HA)[c]	40	7
HA + IS	40	11
Retinoid feeding (RF)[d]	40	47
RF + IS	40	29
RF + HA	40	2
RF + HA + IS	40	0

[a] DMBA was administered at 53 days of age.
[b] IS, cell particulate of DMBA-induced rat mammary carcinomas + Freunds complete adjuvant, administered i.p. at 1, 3, and 5 weeks after DMBA treatment.
[c] HA, tamoxifen + CB-154 treatments, administered s.c. daily commencing 3 days after DMBA treatment.
[d] RF, retinyl acetate feeding (1.0 mM), daily, commencing 3 days after DMBA treatment.

From Welsch, C. W. and DeHoog, J. V., *Cancer Res.*, 43, 585, 1983. With permission.

treated rats, we are aware of only one report that evaluates the effect of this dietary constituent on development of spontaneous mammary carcinomas in this species. In the Lew/Mai rat strain, a relatively high incidence of spontaneous mammary carcinomas arise in animals allowed to live a full life time. McCormick et al.[34] fed female Lew/Mai rats retinyl acetate for 92 weeks, commencing at 5 weeks of age. Mammary carcinoma incidence in retinyl acetate-fed animals was 28%. In contrast, 30% of the control animals developed these tumors. Thus, it appears that dietary retinyl acetate is ineffective in suppressing the genesis of spontaneous rat mammary carcinomas, at least in the Lew/Mai rat strain.

We are aware of only two reports in which the effect of retinoids on growth of rat mammary carcinoma cells was examined in vitro. Retinoic acid or retinyl acetate inhibited growth of mammary tumor cells obtained from Fischer 344 rats (cell lines 13762NF, DMBA#8, and R-3230AC) by 39 to 88%.[38] Retinoic acid was consistently superior to retinyl acetate in inhibiting cellular proliferation of the mammary carcinoma cell lines. Growth inhibition by the retinoids was most pronounced in lines 13762NF and DMBA#8, both derived from female rats treated with DMBA. In the second report, retinol, retinal, retinoic acid, and retinyl acetate were effective in reducing proliferation of a rat mammary carcinoma cell line (Rama 25) isolated from a Sprague-Dawley rat bearing a DMBA-induced rat mammary carcinoma.[39]

B. Mice

Although there have been a number of reports documenting the chemopreventive activity of a variety of dietary retinoids in rat mammary gland tumorigenesis, fewer laboratories have examined this activity in mice. These studies are summarized in Table 4.[40-46] It is clear that the response to dietary retinoids is quite varied ranging from inhibition, to no effect, and even stimulation of mouse mammary tumorigenesis. We should point out that mice are unable to tolerate the high dietary levels of natural retinoids (e.g., retinyl acetate) that are commonly fed to rats (0.6 to 1.0 mM) in chemoprevention studies. Dietary levels of retinyl acetate of 0.2 mM are those which are "maximally tolerated" by this species, i.e., dietary levels of retinyl acetate in excess of this amount cause significant reductions in mouse body

Table 4
**DIETARY RETINOIDS AND THE INCIDENCE OF MAMMARY
TUMORS IN MICE**

Mouse strain	Retinoid	Method of mammary tumor induction	Effect	Ref.
C3H-A^vy - nulliparous	Retinyl acetate	Spontaneous	No effect	40
C3H - nulliparous	Retinyl acetate	Spontaneous	No effect	41
C3H - nulliparous	4-HPR	Spontaneous	Suppression	41
C3H - multiparous	Retinyl acetate	Spontaneous	No effect	41
C3H - multiparous	4-HPR	Spontaneous	No effect	41
GR - nulliparous	Retinyl acetate	Estrogen/progesterone	Stimulation	42
GR - multiparous	Retinyl acetate	Estrogen/progesterone	Stimulation	42
BD2F₁ - nulliparous	Retinyl acetate	DMBA	No effect	43
BD2F₁ - nulliparous	4-HPR	DMBA	No effect	43
C3H/st/Ha	Retinyl palmitate	Spontaneous	Stimulation	44
C3H/HeJ	Retinyl palmitate	Transplantation	No effect[a]	45
C3H/HeJ	β-carotene	Transplantation	Suppression	46

[a] An initial growth inhibition was noted in this study.

weight gains. It is important to recall that this level of dietary retinyl acetate (0.2 mM) is not effective in the chemoprevention of rat mammary gland carcinogenesis.[25] Therefore, the negative data derived from the retinyl acetate studies shown in Table 4, at least in part, could be explained by quantitatively insufficient levels of dietary retinyl acetate. 4-HPR, a synthetic retinoid, can be fed to mice at high dietary levels (e.g., 1.0 mM) without any adverse side effects. When 4-HPR is fed to nulliparous C3H mice, commencing at 2 months of age, a significant reduction in the incidence of spontaneous mammary carcinomas is observed.[41] The lack of effect of dietary 4-HPR in the prophylaxis of mammary tumorigenesis in multiparous C3H mice and in BD2F₁ mice treated with DMBA provides evidence, however, which supports the concept that mouse mammary gland tumorigenesis is less responsive to the inhibitory actions of dietary retinoids than are carcinogen-induced rat mammary carcinomas. Furthermore, it appears that elevated levels of dietary retinyl acetate or retinyl palmitate, under certain experimental conditions, may even enhance this neoplastic process.[42,44]

A number of laboratories have examined the effect of various retinoids in vitro on the induction and progression of mouse mammary gland tumors. Whole mammary glands obtained from certain strains of estradiol and progesterone primed mice can be cultured in a serum-free media enriched with mammotropic hormones. By 10 days of culture, the mammary glands contain fully developed lobulo-alveoli. If a carcinogen is added to the culture media for one day (usually day 4 of culture), nodule-like alveolar lesions (NLAL) are observed in the cultured glands by 20 days postcarcinogen treatment. When NLAL are transplanted to gland-free fat pads of syngeneic hosts, lobular adenocarcinomas often arise in the grafted tissues suggesting that NLAL contain transformed cells.[47]

Sorof and colleagues[48] examined the effects of a synthetic retinoid, 2-retinylidene-5,5-dimethyl-1,3-cyclohexanedione (retinylidene dimedone) on neoplastic transformation of Balb/c mouse mammary gland induced by DMBA. The mammary glands were cultured for 10 days in growth-promoting media (containing mammotropic hormones) and for an additional 14 days in a media that lack mammotropic hormones and cause mammary gland regression. DMBA was added to the culture media for only 1 day, i.e., on day 4. Retinylidene dimedone was added to the culture media at days 0 to 3, 3 to 4, 4 to 10, 7 to 12, or 10 to 24 of culture. The percent of mammary glands containing NLAL (transformed glands) in control cultures ranged from 30 to 45%. The percent of glands containing NLAL in retinylidene dimedone-treated cultures, when the retinoid was added to culture media at 0 to 3, 3 to 4, 4 to 10, 7 to 12, or 10 to 24 days of culture, was 28, 25, 3, 5, 15%, respectively. Thus, significant suppression of in vitro neoplastic transformation of the mouse mammary gland was observed only when the retinoid was administered after the carcinogen. When the retinoid was administered either before or during carcinogen treatment, mammary gland transformation was comparable to control values. In subsequent studies, using different carcinogens [BP,N-2-fluorenylacetamide (FAA), N-methyl-N-nitro-N-nitrosoquanidine (MNNG), N-nitrosodiethylamine (DENA)] and retinoids (retinylidene dimedone, 4-HPR), similar results were obtained, i.e., the retinoids were effective in suppressing the genesis of NLAL when administered after carcinogen treatment.[49-51] Interestingly, retinylidene dimedone treatments were not effective in suppressing the genesis of NLAL when several direct acting carcinogens (i.e., BP-diol, BP-diol epoxide, N-OH-FAA, N-ACO-FAA, or MNU) were used in these studies.[49] Not only is the in vitro development and/or progression of carcinogen-induced NLAL affected by retinoids, but the in vitro development and/or maintenance of spontaneous mouse mammary gland preneoplastic alveolar lesions (MAL) appear to be as well. Telang and Sarkar[52] recently reported that 4-HPR suppressed the development and inhibited maintenance of MAL in organ cultures of mammary glands from R111 mice.

The addition of certain retinoids to be the media of cell cultures of mouse mammary tumor cells has been reported to suppress proliferation of these cells.[38,53] Mouse mammary tumor cell lines in which proliferation is suppressed by retinyl acetate or retinoic acid are Mm5mT cells (from C3H mice), M12 cells (from C3H mice), and DD3 cells (from carcinogen-treated Balb/c mice). Thus, in vitro, retinoids have been reported to not only suppress the progression of incipient mouse mammary tumor cells but, in addition, suppress the proliferation of more advanced mammary tumor cells, i.e., cells from established mouse mammary tumor cell lines.

The effect of retinoids on development and growth of normal mouse mammae has received considerably less attention. Mehta et al.[54] examined the effect of retinoids on hormone-induced differentiation of the Balb/c mouse mammary gland in vitro. Retinoic acid and 4-HPR inhibited prolactin-induced end bud differentiation into alveolar structures. Retinyl acetate did not affect the morphologic prolactin-induced proliferation of the gland. The effect of these retinoids on epithelial DNA synthesis of the mammae was also examined. Retinoic acid, retinyl acetate or 4-HPR inhibited hormone (insulin and prolactin) induced DNA synthesis of epithelium; the retinoids did not affect DNA synthesis of glands in media lacking mammotropic hormones. Paradoxically, retinyl acetate enhanced DNA synthesis of the mammary gland in hormone supplemented media when the retinoid was added to culture media at reduced concentrations. In contrast, Telang and Sarker[52] could not detect any significant effect of 4-HPR on hormone- (insulin, prolactin, aldosterone, and hydrocortisone) induced alveolar proliferation of organ cultures of R111 or C57BL mouse mammary glands.

C. Humans

Only a few studies have evaluated the effects of retinoids on human mammary epithelial cells. In 1979, Lotan[55] examined the effects of retinoic acid on proliferation of human mammary carcinoma cell lines MDA-MB-157, 734B, SK-BR-3, Hs578T, and noncancerous human mammary cell lines Hs578Bst and HBL-100. Growth of mammary carcinoma cell lines SK-BR-3 and 734B was significantly inhibited by retinoic acid; growth of cell line Hs578T was only slightly inhibited while the proliferation of cell line MDA-MB-157 was not affected. Proliferation of the noncancerous mammary cell line HBL-100 was slightly inhibited and that of Hs578Bst was not affected. Further studies with the more sensitive of these cell lines revealed that the inhibitory effects of retinoic acid were dose dependent.

Ueda et al.[56] reported the effects of various retinoids on proliferation of human mammary carcinoma cells (MCF-7) in vitro. Retinoids that significantly decreased cell proliferation of MCF-7 were retinoic acid, retinal, retinol, retinyl palmitate, and retinyl acetate. Retinoic acid revealed maximum activity in inhibiting cell proliferation. More recently, it was reported that retinoic acid is more effective than retinol in reducing proliferation of human breast cancer cells (MCF-7, Hs578T, ZR-75-B) in vitro.[57]

To date, there has been only one report describing the therapeutic potential of retinoids in control of *in situ* growth of human breast carcinoma. Cassidy et al.[58] examined the activity of 13-*cis*-retinoic acid in 18 patients with advanced breast cancer refractory to standard cytotoxic and/or endocrine therapy. Patients were first treated with 13-*cis*-retinoic acid, and then the level of the retinoid was escalated over a 1-month period, unless toxicity (dry skin, dry mucosa, cheilitis, conjunctivitis) forced dose reduction. Four patients exhibited hypercalcemia, two complained of severe earache, and several had nausea, vomiting, and abdominal cramping. There were no objective tumor responses as defined by standard criteria.

D. Synopsis

Under near identical experimental conditions, dietary retinoids can effectively suppress the genesis of spontaneous mammary tumors in nulliparous C3H mice but not in multiparous C3H mice. Dietary retinoids can also effectively suppress the genesis of mammary tumors in carcinogen-treated rats but not in carcinogen-treated mice. In these studies, the only apparent relationship between rodent models that respond to retinoids and those which are non responders is the stage of neoplastic development and/or progression at onset of retinoid treatment. Clearly, the multiparous C3H mouse (mammary glands contain numerous preneoplastic lesions) has progressed considerably farther than the young nulliparous C3H mouse (free of overt preneoplastic mammary lesions) with regard to mammae neoplastic potential. Carcinogen-induced mammary tumors in mice (induced by multiple carcinogen injections) are considerably more anaplastic (less differentiated, respond less to endogeneous growth factors) than carcinogen-induced mammary tumors in rats (induced by a single carcinogen injection). In accord, in vitro studies provide evidence that retinoids are ineffective in suppressing the progression of carcinogen-induced mouse mammary NLAL when NLAL are induced by potent direct acting carcinogens; less potent carcinogens (those which require metabolic activation) induce NLAL whose progression is inhibited by retinoids.

A similar relationship can be found in yet another experimental system. Retinoids are effective in suppressing the promoting action of 12-*O*-tetradecanoylphorbol-13-acetate (TPA) on carcinogen-induced (single carcinogen injection) skin tumors in mice but are ineffective in suppressing the genesis of the more anaplastic skin tumors induced by multiple carcinogen injections.[59] These results suggest that the efficacy of retinoids is enhanced when the target neoplastic cells are less anaplastic, i.e., more akin physiologically to their tissue of origin. The failure to achieve suppression of human breast carcinoma progression with retinoid treatment, as previously described, may be, at least in part due to the relatively late stage of tumor development upon onset of retinoid therapy. Although several cell lines of rodent mammary tumors, all of which are highly anaplastic, did respond to retinoid treatment in vitro, the concentrations of the retinoids in the culture media (frequently $10^{-5}M$) were almost invariably many orders of magnitude higher than those found in serum ($10^{-6}M$ retinol or $10^{-9}M$ retinoic acid).

III. MECHANISM BY WHICH RETINOIDS MAY INFLUENCE MAMMARY GLAND TUMORIGENESIS

A. Inhibition of Mammary Tumorigenesis by Retinoids: At the Initiation or Promotional Level?

Temporal studies have revealed a common feature in the mechanism of action of retinoids in the prophylaxis of chemical carcinogenesis of the rat mammary gland. Retinoids (retinyl acetate, retinyl methyl ether, 4-HPR) can effectively suppress the genesis of carcinogen-(DMBA, MNU, BP) induced rat mammary carcinomas when the retinoid is administered to animals commencing at various time periods *after* carcinogen treatment.[15,22-34] However, its continued presence is necessary for effective chemoprevention; upon retinoid withdrawal, a surge in mammary tumor incidence is observed.[29,60] Thus, it is clear that dietary retinoids can act as potent antipromoters in chemical carcinogenesis of the rat mammary gland. There is evidence, however, that the antitumor promoting activity of retinoids is reduced when retinoid feeding is commenced only during the later stages of tumor development.[35]

Of equal significance are the observations that dietary retinoids (i.e., retinyl acetate) can effectively suppress rat mammary gland chemical carcinogenesis when the retinoid is administered before and during carcinogen (DMBA, BP) treatment.[22,34] Whether or not retinyl acetate-induced hepatic toxicity is a factor (e.g., alteration of carcinogen metabolism) in these studies is not known. Nevertheless, it is clear that retinyl acetate is an effective inhibitor of initiation of rat mammary gland carcinogenesis. These studies provide support for the concept that retinoids can effectively inhibit rat mammary gland carcinogenesis at two stages of tumorigenesis, i.e., initiation and promotion.

Only two laboratories have reported that dietary retinoids (4-HPR and β-carotene) can suppress the genesis of mammary tumors in mice. Dietary 4-HPR significantly reduced the incidence of spontaneous mammary tumors in nulliparous C3H mice[41] and dietary β-carotene slowed the growth of transplanted mammary carcinomas in C3H/HeJ mice.[46] The latter study supports the concept that dietary retinoids can affect the promotional stage of mouse mammary gland tumorigenesis; neither study provides insight into whether or not retinoids affect the initiation stage of this neoplastic process.

The effect of retinoids on chemical carcinogenesis of the mouse mammary gland has been examined in vitro for the ultimate purpose of determining whether or not retinoids can affect either the initiating or promoting stages of this carcinogenic process. Significant suppression of in vitro neoplastic transformation of the mouse mammary gland was observed only when the retinoid (retinylidene dimedone) was administered after the carcinogen (DMBA). When the retinoid was administered either before or during carcinogen treatment, mammary gland transformation was comparable to control values.[48,49] Using a different carcinogen (DENA) and a different retinoid (4-HPR), in vitro inhibition of mouse mammary gland transformation was reported when the retinoid was administered after carcinogen treatment.[50,51]

In conclusion, it appears that retinoids can inhibit the promotional phase of rat and mouse mammary gland carcinogenesis as well as the initiating phase of rat mammary gland carcinogenesis. Suppression by retinoids of initiating events of mouse mammary gland carcinogenesis has not yet been described.

B. Mechanisms
1. Initiation Stage
a. Modifications of Carcinogen Metabolism

DMBA and a number of other polycyclic aromatic hydrocarbons exert their mutagenic and carcinogenic effects only after metabolic activation, which produces ultimate carcinogens that bind covalently to biologically active macromolecules such as DNA, RNA, and protein. Several studies have demonstrated that retinoids inhibit the binding of carcinogen to DNA of a variety of tissues[61,62] including the mammary gland,[63] while vitamin A deficiency enhances such binding.[64,65] The mechanism by which retinoids affect carcinogen binding is not known.

A possible mechanism to explain the inhibition by vitamin A of carcinogen binding to DNA involves competition with the activated or ultimate carcinogen. Vitamin A readily forms epoxides. Since ultimate carcinogens may be epoxides, vitamin A could present itself as another substrate for epoxide generating systems. It is well known that 5,6-epoxyretinoic acid is formed in tissues in vivo.[66] Thus, it is possible that epoxidation of vitamin A compounds may be one of the biological mechanisms that interfere with carcinogen activation. Alternatively, the inhibiting effect of retinoids on carcinogen binding to DNA may be mediated via prostaglandins since prostaglandin production can be enhanced by vitamin A compounds[67,68] and prostaglandins have been reported to inhibit carcinogen binding to DNA.[69]

Several studies have demonstrated that retinoids can directly modulate metabolic activation of carcinogens that are catalyzed by microsomal monoxygenases such as aryl hydrocarbon hydroxylase. For example, retinoids were found to inhibit the induction of BP-activating enzymes in mice,[70] while in vitamin A-deficient guinea pig microsomes the activity of carcinogen-activating enzymes was high.[71] These studies provide evidence that retinoids can prevent the formation of the ultimate carcinogen.

It is clear that the anticarcinogenic actions of retinoids at initiation cannot be entirely explained by an intervention in carcinogen activation, since retinoids can prevent tumor (colon) induction by carcinogens that do not require metabolic activation.[72] Thus, additional mechanisms must be postulated for the prevention of initiation by retinoids of many carcinogens. Several studies have demonstrated that hepatic degradation of carcinogens is decreased in vitamin A-deficient animals, suggesting that vitamin A may enhance carcinogen inactivation.[73,74]

b. Modification of Cell Division

Any substance which inhibits normal mammary gland cell division would, in addition, have the capability of suppressing the initiating activities of chemical carcinogens for it is generally accepted that replicating cells are more vulnerable to carcinogen action than are nonreplicating cells. In general, treatments that suppress normal rat mammary gland DNA synthesis when implemented before and during carcinogen treatment, reduce the incidence of mammary carcinomas.[74,75] If it can be shown that retinoids inhibit mammary gland cell division in vivo (or in vitro), an attractive hypothesis explaining the inhibitory effects of dietary retinoids at initiation would be provided.

Mehta and Moon[77] reported that the mammae of carcinogen- (DMBA and MNU) treated female Sprague-Dawley rats fed retinyl acetate had lower DNA synthetic activity than placebo-fed controls, as judged by ^{3}H-thymidine incorporation into chemically extracted DNA. The interpretation of these data is difficult because the retinyl acetate-induced inhibition of DNA synthesis occurred only in carcinogen-treated rats; in rats not treated with carcinogens, retinyl acetate did not affect mammary gland DNA synthesis. McCormick et al.[34] reported that female Lew/Mai rats fed retinyl acetate had decreased DNA synthesis in the mammae judged by radioautographic analysis of ^{3}H-thymidine uptake by mammary gland epithelium. Morphological assessment of these glands, however, revealed only a slight reduction in mammae development. Welsch et al.[25,29,36] and Moon et al.[33] have examined, by whole-mount inspection, mammary glands from rats treated with pharmacological levels of retinoids and have observed reduced cellular proliferation or altered morphology (reduced number of ducts and alveoli) of the epithelium within these glands. This has been observed in rats fed either retinyl acetate or 4-HPR. Thus, it appears that pharmacological levels of retinoids are antiproliferative to the normal rodent mammary gland in vivo although this question has not been definitively answered. Interestingly, a recent study provides evidence that vitamin A is not required for normal growth and differentiation of the mammary gland.[78] Mammary glands from female Sprague-Dawley rats that are severely deficient in vitamin A have as much epithelium as glands from control animals. Furthermore, explants from mammary glands of vitamin A-deficient rats were as responsive to mammotrophic hormones as explants from control animals. These very interesting and provocative findings suggest that moderate amounts of vitamin A do not have a physiological role either in the maintenance of rat mammary epithelium or in the alteration of its potential for hormone-dependent phenotypic expression of differentiation.

2. Promotion Stage

a. Retinoid-Binding Proteins

Retinoids may act through cellular binding proteins in a manner similar to that described for steroid hormones. A cellular protein which specifically binds retinol has been detected and characterized. A second protein which specifically binds retinoic acid has also been detected and has been partially characterized. It has been proposed that the retinoid-binding protein complex migrates from cytoplasm to nucleus where it influences gene expression.[79-84]

Cellular retinoic acid-binding protein (CRABP) in mammary tissues was first described by Ong et al.[81] in 1975. Extracts of human breast carcinomas were found to contain CRABP, whereas, in contrast, this protein was not detected in grossly normal breast tissues adjacent to tumor, obtained from the same patients. Similar results were reported by Huber et al.[85] Utilizing a more sensitive technique for the determination of CRABP, Kung et al.[86] examined 88 human breast specimens of differing pathologies for the presence of this protein. CRABP was found in all breast specimens examined with progressively increasing amounts from normal to carcinomatous tissues. CRABP has also been shown in a human breast cancer cell line (MCF-7).[57,87] No significant correlation between the presence or absence of steroid hormone receptors and CRABP in the cytosols of human breast carcinomas has been reported.[88]

First evidence of retinoid-binding proteins in rodent mammary tissues was obtained by Ong and Cytil[89] in 1976. Cellular retinol-binding proteins (CRBP) and CRABP were observed in two transplanted mammary carcinoma cell lines, i.e., the Walker 256 carcinosarcoma maintained in Sprague-Dawley rats and MAC-1 carcinoma maintained in Fischer 344 rats. Subsequently, CRABP and/or CRBP have been reported in normal mammary glands of virgin, pregnant, and lactating Sprague-Dawley rats and virgin C57BL mice and in C3H mouse mammary tumors and carcinogen-induced rat mammary tumors.[90-94] In carcinogen-induced rat mammary carcinomas, CRABP has been reported to be found in nuclear and cytosolic cellular fractions.[84] CRABP concentrations are lower in the normal rat mammary gland when compared with carcinomatous mammae.[92] The administration of estrogen or estrogen with progesterone to virgin or lactating Sprague-Dawley rats reportedly increases mammary gland concentrations of CRABP.[94] CRABP concentrations appear to be lower in hormone-dependent rat mammary carcinomas than in tumors that are hormone independent.[93] Although experiments have demonstrated that CRABP or CRBP is present in higher concentrations in neoplastic mammae than in normal mammary tissue, this difference may reflect the greater amount of extracellular stroma found in normal vs. neoplastic mammary tissues. This problem was recently addressed by Bunk et al.[95] who reported higher concentrations of CRABP and CRBP in cell cultures of mouse mammary tumors (C57BLfR111) than in cell cultures of normal mouse (C57BL) mammae. Thus, it appears that the higher concentrations of CRABP and CRBP found in neoplastic mammae actually reflect increased intracellular levels of retinoid-binding proteins. It remains to be determined whether or not the intracellular metabolic effects of retinoids on normal or neoplastic mammary gland are exerted partially, primarily, or solely via specific receptor proteins.

Studies of Lotan et al.[96] and Libby and Bertram[97] suggest that retinoids may act via a mechanism that does not involve specific binding proteins as they were unable to detect retinoid-binding proteins in cell lines which were responsive to retinoids. Thus CRBP or CRABP are undetectable in mouse L1210-A5 leukemia cells yet the proliferation of this cell line in vitro is inhibited by retinyl acetate and retinoic acid.[96] Similarly, CRBP and CRABP are undetectable in mouse C3H/10T$^1/_2$ cells yet the transformation of this cell line is inhibited by a variety of natural and synthetic retinoids, including retinyl acetate and retinol.[97] Similarly, CRBP was not detected in the human breast cancer cell line MCF-7,[87] yet retinol was effective in inhibiting growth of this cell line in vitro.[57] In some cases CRBP and CRABP concentrations do change during post- and perinatal growth and development.[98,99] Therefore, these binding proteins may be the cause of or the consequence of a changing differentiated state.

b. Membrane Alterations

It is well known that when vitamin A is added in excess to biological systems, it can act on cellular and subcellular membranes. Although this phenomenon has not been examined in normal or neoplastic mammary gland cells, it has been studied in many other biological and experimental systems. High, nonphysiological concentrations of vitamin A can labilize membranes causing leakage of intracellular components and finally, disruption of tissue.[100,101] A prominent hypothesis for the mechanism of the antipromotional activity of pharmacological doses of retinoids and the resultant release of hydrolytic enzymes accounts for the destruction of neoplastic cells.[102] However, in several studies, no correlation was observed between the ability of compounds to stabilize or labilize lysosomal membranes and their antitumorigenic properties.[18] It is probable, therefore, that labilization of membranes by retinoids is not a valid mechanism to explain the antipromotional activities of retinoids.

Considerable evidence, however, does point to membranes as a possible cellular site for the initiation of physiological as well as the antitumorigenic functions of vitamin A. Among the mechanisms proposed for the physiological function of vitamin A, the role of vitamin A in glycosylation of membrane proteins and lipids is currently being extensively examined.[103] Membrane glycosylation is associated with cell-surface recognition mechanisms that are crucial in the regulation of cellular growth and differentiation. Profound changes in cell surface properties, in addition to cytoplasmic changes, are also linked to neoplasia. Current evidence supports the concept that cell membrane glycoconjugates have an important role in transformation.[104]

Since retinoids, when added to cells in culture, almost invariably alter cell-surface associated phenomena, it is logical to seek an explanation for the antitumorigenic properties of these compounds by examining biochemical alterations in the membranes of carcinogen-initiated transformed cells that have been exposed to retinoids. Modulation of cell surface functions (adhesion, spreading, ion transport, cell-cell communication) and composition (glycoproteins, glycolipids, phospholipids) by retinoids have been studied in numerous in vitro systems. Such studies have demonstrated that retinoids increase cellular adhesion and restore anchorage-dependent growth to transformed cells, possibly by an effect on cell surface protein glycosylation. It is possible that the restoration of normal function (e.g., anchorage-dependent growth, active cell-cell communication, etc.) to a transformed cell may be the basis for the antitumorigenic effects of vitamin A compounds.[18,103-112]

It should be pointed out that this postulated mechanism of retinoid action, i.e., modification of membrane glycosylation may or may not require binding of the retinoid to a specific receptor. Other actions of retinoids (reviewed in Reference 113) on cellular or subcellular membranes which could be important in the antipromotional activity of these compounds are alteration of conformation and/or charge of membrane proteins and/or alteration of membrane electron transfer; these activities may or may not involve a specific receptor-mediated event.

c. Modification of Cell Division
i. Inhibition of the Action of Mitogenic Peptides

An intriguing hypothesis, proposed by Todaro and co-workers,[114] is that retinoids may modulate the mitogenic effects of peptide hormones on specific epithelial target organ sites. This hypothesis was derived, at least in part, from the observation that retinoids can antag-

Table 5
EFFECT OF RETINYL ACETATE FEEDING ON THE INCIDENCE OF MNU-INDUCED MAMMARY CARCINOMAS IN RATS TREATED WITH A PROLACTIN SECRETION STIMULATING DRUG (HALOPERIDOL)

			Total number of palpable mammary carcinomas			
Group	Treatment[a]	Number of rats	Onset of treatment	2 weeks after treatment	4 weeks after treatment	Mean number of mammary carcinomas/rat (4 weeks after treatment)
I	Haloperidol treatment	15	2	17[b]	36[b]	2.4[b]
II	Haloperidol treatment + retinyl acetate feeding	16	1	4[c]	7[c]	0.4[c]

[a] MNU was administered at 50-53 days of age and at 57-60 days of age. Upon the first evidence of palpable mammary tumors, each of 31 rats were treated daily with haloperidol; 16 of these rats were fed daily, in addition, retinyl acetate.

[b,c] $P < 0.01$.

From Welsch, C. W., De Hoog, J. V., Scieszka, K. M., and Aylsworth, C. F., *Cancer Res.*, 44, 166, 1984. With permission.

onize the stimulatory effect of sarcoma growth factor (a peptide) on cell proliferation and anchorage-independent growth of fibroblasts in culture. Recent results from our laboratory provide evidence in support of this hypothesis, extending this concept to mammary gland carcinomas. Prolactin, a peptide hormone, is a well-known rat mammary carcinoma mitogen in vivo.[115] The feeding of retinyl acetate to rats treated with a drug (haloperidol) to induce hyperprolactinemia completely blocked the stimulatory effect of prolactin on mammary carcinoma development (Table 5).[36] Insulin, also a peptide hormone, is a potent mitogen of rat mammary carcinomas in vitro. It has been reported by a number of laboratories that the addition of insulin to cell or organ cultures of rodent mammary gland carcinomas results in a striking increase in ^{3}H-thymidine incorporation into chemically extractable DNA, an increase in ^{3}H-thymidine labeled cells, and an increase in number of cells with mitotic figures.[116] Recent studies in our laboratory demonstrate that the addition of retinoic acid to media of organ cultures of rat mammary gland carcinomas blocks the stimulatory effect of insulin on ^{3}H-thymidine incorporation into DNA (Table 6).[36] Retinoic acid added to the culture media alone did not effect ^{3}H-thymidine incorporation into DNA. Thus, we provide both in vivo and in vitro evidence that retinoids can inhibit the proliferative actions of peptide hormones on rat mammary gland carcinomas. Our results and those of Todaro et al.[114] are similar in many respects to a recent report by Mehta et al.[54] showing inhibition by retinoic acid of prolactin-induced DNA synthesis and lobulo-alveolar development of organ cultures of normal mouse mammae. It is important to point out, however, that in other in vitro systems (e.g., skin fibroblasts), the mitogenic activity of certain peptide growth factors (e.g., EGF, FGF, vasopressin, insulin) appears to be enhanced by vitamin A compounds.[117,118]

ii. Induction of Differentiation: a cAMP Mediated Event?
It is well known that vitamin A active compounds modulate differentiation in many mammalian tissues,[6] although the possibility has been raised that the mammary gland may be an exception.[78] Vitamin A compounds can also enhance differentiation of carcinomatous tissues. Perhaps the most striking example of this phenomenon is the conversion of murine

Table 6
EFFECT OF RETINOIC ACID AND/ OR INSULIN TREATMENTS ON MEAN ^{3}H-THYMIDINE INCORPORATION INTO DNA OF 2-DAY ORGAN CULTURES OF 7 MNU-INDUCED RAT MAMMARY CARCINOMAS

Treatment	Mean ^{3}H-thymidine/μg DNA cpm ($\pm$ SE)
Control	12.8 ± 0.8[a]
Insulin (I)	19.9 ± 3.2[b]
Retinoic acid (RA)	11.9 ± 2.6[a]
I + RA	10.1 ± 1.8[a]

[a,b] $p < 0.05$.

From Welsch, C. W., DeHoog, J. V., Scieszka, K. M., and Aylsworth, C. F., *Cancer Res*, 44, 166, 1984. With permission.

embryonal carcinomas to benign teratomas by retinoids in vivo.[119] Strain 129 mice bearing PCC4 embryonal carcinoma subcutaneously were treated daily (intratumor) with retinoic acid. The retinoid induced nearly complete morphological differentiation mainly into neuroepithelial and glandular derivatives. Differentiation was associated with a decreased tumor growth rate, decreased mitotic index, and increased survival time of hosts. In 4 of 18 cases, long-term survival of the hosts was effected by complete differentiation of the malignant embryonal carcinoma tumors into benign teratomas. This very interesting study provides evidence that retinoids cannot only induce differentiation in normal epithelium but also in carcinomatous tissues. Whether such a phenomenon occurs in rat mammary carcinomas as a result of retinoid feeding is not known. Histological assessment of mammary tumors from rats treated with pharmacological levels of retinoids provides no indication of retinoid-induced tumor cell differentiation, i.e., no morphological evidence of secretion in the tumor cells is observed.[25,29,36] Furthermore, the morphological examination of the normal mammae of these animals has not provided any indication of enhanced differentiation (secretion and/ or alveoli formation). In contrast, when GR mice are fed large doses of retinyl acetate, a marked increase in mammary alveoli development is observed in these animals.[42] This observation is consistent with the recent report by Rudland et al.[39] showing enhanced differentiation (casein and dome formation) of rat mammary carcinoma cells (Rama 25) by retinoids in vitro.

Studies on differentiation (and proliferation) of a variety of cell types, including normal and neoplastic mammae[120,121] have focused on cyclic adenosine 3′,5′-monophosphate (cAMP) as a mediator for these events. Embryonal carcinoma F9 cells differentiate in vitro into parietal endoderm in the presence of retinoic acid.[122,123] Differentiation of these cells in vitro can also be accomplished by increasing cAMP levels. Subsequent cell culture studies have shown that vitamin A compounds increase cAMP-dependent protein kinase activity in several cell types with concurrent enhancement of differentiation.[124-126] Vitamin A compounds may affect the regulatory subunits (R_1 and R_2) of this enzyme in cytosolic and plasma membrane fractions.[126] Retinoic acid has been reported to increase two proteins of molecular weights of 160 and 250 kd that undergo cAMP-dependent phosphorylation.[127] These studies suggest that vitamin A compounds increase the activity of cAMP-dependent protein kinase, an event that might alter differentiation and/or proliferation of cells. These studies did not examine cells of the mammae. The only report to our knowledge examining the relationship between cAMP, retinoid treatment, and mammary tumorigenesis in rodents was that by Welsch and DeHoog.[128] Retinyl acetate feeding blocked the stimulatory effect of cholera toxin (increases intracellular cAMP levels) on the development of MNU-induced rat mammary carcinomas.

The concept that retinoids can exert their antipromotion effect on tumor cells via a mechanism which enhances cell differentiation is an intriguing concept. This concept receives additional support from in vitro data obtained with human promyelocytic leukemia (HL-60) cells in which retinoic acid was shown to stimulate differentiation of these cells as well as suppressing *myc* oncogene activity.[129] Furthermore, retinoids have been reported to inhibit proliferation and enhance differentiation of human neuroblastoma cells (LA-N-1) in vitro.[130]

d. Immune Stimulation

There is increasing evidence that retinoids are immunoregulatory.[131-133] Vitamin A has been reported to be an immune adjuvant.[134] For example, the administration of certain retinoids to laboratory animals has been reported to cause thymus and lymph node enlargement,[135] to enhance humoral antibody response to a variety of antigens,[134] to enhance certain cell-mediated responses,[136] and to enhance tumoricidal macrophage activity.[137]

We are aware of only three studies that examined the interaction of retinoid feeding and immune stimulation on the genesis of murine mammary tumors.[25,36,37] In all three studies, pharmacological levels of retinoids were used. Enhancement of retinyl acetate activity in the suppression of DMBA- or MNU-induced rat mammary gland carcinogenesis by concurrent treatment with an immunostimulant consisting of pooled cell particulates of carcinogen-induced rat mammary carcinomas (c̄ Freund's complete adjuvant) has been reported.[25,36] Other immunostimulants (e.g., BCG, *Norcardia rubra*, MVE-2) were not effective in enhancing the chemopreventive activities of retinoids in this carcinogenic process.[25,37] A significant synergism between retinoid treatment and various immune modifying therapies has been reported for carcinogen-induced murine fibrosarcomas.[138,139] It should be pointed out, however, that in the studies showing a synergism between immune stimulation and retinoid treatment, it was not established that this synergism was a result of enhanced immune system reactivity.

The concept that the antitumorigenic effects of pharmacological levels of retinoids is via the immune system is attractive. Nevertheless, it is yet to be demonstrated that retinoids enhance immune effector cells or noncellular immune activities which are specifically directed toward neoplastic mammary epithelium. Furthermore, the ability of retinoids to exhibit antitumorigenic effects in vitro, where immune processes are absent, provides evidence that immune stimulation is not the only mechanism by which retinoids may influence this tumorigenic process.

IV. CONCLUSIONS

The effect of dietary retinoids in chemoprevention of carcinogen-induced rat mammary carcinomas has been studied by several laboratories. Clearly, natural and certain synthetic retinoids are efficacious inhibitors of this neoplastic process, when the retinoids are fed either before or after carcinogen treatment. Only in one other system has inhibition of *in situ* mammary tumorigenesis by dietary retinoids been demonstrated, i.e., in the nulliparous C3H mouse model. Dietary retinoids have not been effective in suppressing mammary tumorigenesis in five models, i.e., the nulliparous C3H-A[vy] mouse, the multiparous C3H mouse, the hormone-treated GR mouse, the carcinogen-treated BD2F₁ mouse, and spontaneous mammary carcinomas in Lewis rats. Only one study to date has examined the effect of retinoids (13-*cis*-retinoic acid) on the progression of breast cancer in humans; no effect of this retinoid was observed.

In the only experimental mammary tumor system in which dietary retinoids are consistently effective (i.e., the carcinogen-induced rat mammary carcinoma model), substantial amounts of the retinoid in the diet (0.6 to 1.0 mM) are required before a significant chemopreventive effect can be observed. A requirement for such large amounts of retinoids in the diet plus the observation that a striking number of mammary tumor models are nonresponsive to this dietary component (as well as the fact that some mammary tumor models are stimulated by retinoids) most assuredly tempers ones enthusiasm for the potential application of retinoids in the control of the human disease.

Nevertheless, many of the shortcomings cited above could conceivably be overcome by the use of synthetic retinoids with enhanced biological activity and reduced toxicity; research efforts in this direction should be strongly encouraged. Increased efforts should also be directed toward the study of the mechanisms by which retinoids promote differentiation and/ or reduce proliferation of neoplastic cells. Such research efforts should not only reveal the biochemical actions of retinoids but in addition should provide new insight into the mechanisms of tumor cell proliferation. Although many biologically active mitogenic stimulating factors are available for laboratory study, few growth-inhibiting compounds have been described; the retinoids are among these compounds and herein lies their primary importance.

ACKNOWLEDGMENTS

Supported by research grant NIH-CP-05717 to C. Welsch and research grant NIH-CA-33190 to M. Zile and C. Welsch.

REFERENCES

1. **Stepp, W.,** Versuche uber fulterung mit lipoidfreier nahrung, *Biochem. Z.,* 22, 452, 1909.
2. **Drummond, J. C.,** The nomenclature of the so-called accessory food factors (vitamins), *Biochem. J.,* 14, 660, 1920.
3. **Karrer, P., Morf, R., and Schopp, K.,** Zur kenntnis des vitamins-A aus fischtranen, *Helv. Chim. Acta,* 14, 1036, 1931.
4. **Baxter, J. C. and Robeson, C. D.,** Crystalline vitamin A, *J. Am. Chem. Soc.,* 64, 2411, 1942.
5. **Isler, O., Ronco, A., Guex, W., Hindley, N. C., Huber, W., Drater, K., and Kofler, M.,** Uber die ester und ather des synthetischen vitamins A, *Helv. Chim. Acta,* 32, 489, 1949.
6. **Wolbach, S. B. and Howe, P. R.,** Tissue changes following deprivation of fat soluble vitamin A, *J. Exp. Med.,* 42, 753, 1925.
7. **Fujimaki, Y.,** Formation of carcinoma in albino rats fed on deficient diets, *J. Cancer Res.,* 10, 469, 1926.
8. **Becci, P. J., Thompson, H. J., Grubbs, C. J., Squire, R. A., Brown, C. C., Sporn, M. B., and Moon, R. C.,** Inhibitory effect of 13-cis-retinoic acid on urinary bladder carcinogenesis induced in C57BL/ 6 mice by N-butyl-N-(4-hydroxybutyl)-nitrosamine, *Cancer Res.,* 38, 4463, 1978.
9. **Chu, E. W. and Malmgren, R. A.,** An inhibitory effect on vitamin A on the induction of tumors of forestomach and cervix in the Syrian hamster by carcinogenic polycyclic hydrocarbons, *Cancer Res.,* 25, 884, 1965.
10. **Newberne, P. M. and Suphakarn, V.,** Preventive role of vitamin A in colon carcinogenesis, *Cancer,* 40, 2553, 1977.
11. **Bollag, W.,** Therapy of chemically induced skin tumors of mice with vitamin A palmitate and vitamin A acid, *Experientia,* 27, 90, 1971.
12. **Saffiotti, U., Montesano, R., Sellakumar, A. R., and Borg, S. A.,** Inhibition by vitamin A of the induction of tracheobronchial squamous metaplasia and squamous cell tumors, *Cancer,* 20, 857, 1967.
13. **Longnecker, D. S., Kuhlman, E. T., and Curphey, T. J.,** Divergent effects of retinoids on pancreatic and liver carcinogenesis in azaserine-treated rats, *Cancer Res.,* 43, 3219, 1983.
14. **Daoud, A. H. and Griffin, A. C.,** Effect of retinoic acid, butylated hydroxyltoluene, selenium and sorbic acid on azo-dye hepatocarcinogenesis, *Cancer Lett.,* 9, 299, 1980.
15. **Moon, R. C., Grubbs, C. J., and Sporn, M. B.,** Inhibition of 7,12-dimethylbenzanthracene-induced mammary carcinogenesis by retinyl acetate, *Cancer Res.,* 36, 2626, 1976.

16. **Sporn, M. B. and Newton, D. L.**, Chemoprevention of cancer with retinoids, *Fed. Proc.*, 38, 2528, 1979.
17. **Nettesheim, P.**, Inhibition of carcinogenesis by retinoids, *Can. Med. J.*, 122, 757, 1980.
18. **Lotan, R.**, Effects of vitamin A and its analogues (retinoids) on normal and neoplastic cells, *Biochem. Biophys. Acta*, 605, 33, 1980.
19. **Newberne, P. M. and Rogers, A. E.**, Vitamin A, retinoids and cancer, in *Nutrition and Cancer: Etiology and Treatment*, Newell, G. R. and Ellison, N. M., Eds., Raven Press, New York, 1981, 217.
20. **Verma, A. K.**, Biochemical mechanisms of modulation of skin carcinogenesis by retinoids, in *Retinoids*, Orfanos, C. E., Ed., Springer-Verlag, New York, 1981, 117.
21. **Bollag, W.**, Retinoids and cancer, *Cancer Chemother. Pharmacol.*, 3, 207, 1979.
22. **McCormick, D. L., Burns, F. J., and Albert, R. E.**, Inhibition of rat mammary carcinogenesis by short dietary exposures to retinyl acetate, *Cancer Res.*, 40, 1140, 1980.
23. **Ip, C. and Ip, M. M.**, Chemoprevention of mammary tumorigenesis by a combined regimen of selenium and vitamin A, *Carcinogenesis*, 2, 915, 1981.
24. **Thompson, H. J., Meeker, L. D., Tagliaferro, A. R., and Becci, P. J.**, Effect of retinyl acetate on the occurrence of ovarian hormone-responsive and -nonresponsive mammary cancers in the rat, *Cancer Res.*, 42, 903, 1982.
25. **Welsch, C. W. and DeHoog, J. V.**, Retinoid feeding, hormone inhibition, and/or immune stimulation and the genesis of carcinogen-induced rat mammary carcinomas, *Cancer Res.*, 43, 585, 1983.
26. **Grubbs, C. J., Moon, R. C., Sporn, M. B., and Newton, D. L.**, Inhibition of mammary cancer by retinyl methyl ether, *Cancer Res.*, 37, 599, 1977.
27. **McCormick, D. L., Mehta, R. G., Thompson, C. A., Dinger, N., Caldwell, J. A., and Moon, R. C.**, Enhanced inhibition of mammary carcinogenesis by combined treatment with N-(4-hydroxy-phenyl)retinamide and ovariectomy, *Cancer Res.*, 42, 508, 1982.
28. **Moon, R. C., Grubbs, C. J., Sporn, M. B., and Goodman, D. G.**, Retinyl acetate inhibits mammary carcinogenesis induced by N-methyl-N-nitrosourea, *Nature (London)*, 267, 620, 1977.
29. **Welsch, C. W., Brown, C. K., Goodrich-Smith, M., Chiusano, J., and Moon, R. C.**, Synergistic effect of chronic prolactin suppression and retinoid treatment in the prophylaxis of N-methyl-N-nitrosourea-induced mammary tumorigenesis in female Sprague-Dawley rats, *Cancer Res.*, 40, 3095, 1980.
30. **Thompson, H. J., Meeker, L. D., and Becci, P. J.**, Effect of combined selenium and retinyl acetate treatment on mammary carcinogenesis, *Cancer Res.*, 41, 1413, 1981.
31. **McCormick, D. L., Sowell, Z. L., Thompson, C. A., and Moon, R. C.**, Inhibition by retinoid and ovariectomy of additional primary malignancies in rats following surgical removal of the first mammary cancer, *Cancer*, 51, 594, 1983.
32. **Moon, R. C. and McCormick, D. L.**, Inhibition of chemical carcinogenesis by retinoids, *J. Am. Acad. Dermatol.*, 6, 809, 1982.
33. **Moon, R. C., Thompson, H. J., Becci, P. J., Grubbs, C. J., Gander, R. J., Newton, D. L., Smith, J. M., Phillips, S. L., Henderson, W. R., Mullen, L. T., Brown, C. C., and Sporn, M. B.**, N-(4-hydroxyphenyl)retinamide, a new retinoid for prevention of breast cancer in the rat, *Cancer Res.*, 39, 1339, 1979.
34. **McCormick, D. L., Burns, F. J., and Albert, R. E.**, Inhibition of benzopyrene-induced mammary carcinogenesis by retinyl acetate, *J. Natl. Cancer Inst.*, 66, 559, 1981.
35. **Gandilhon, P., Melancon, R., Djiane, J., and Kelly, P. A.**, Comparison of ovariectomy and retinyl acetate on the growth of established 7,12-dimethylbenzanthracene-induced mammary tumors in the rat, *J. Natl. Cancer Inst.*, 69, 447, 1982.
36. **Welsch, C. W., DeHoog, J. V., Scieszka, K. M., and Aylsworth, C. F.**, Retinoid feeding, hormone inhibition and/or immune stimulation and the progression of MNU-induced rat mammary carcinoma: suppression by retinoids of peptide hormone induced tumor cell proliferation in vivo and in vitro, *Cancer Res.*, 44, 166, 1984.
37. **McCormick, D. L., Becci, P. J., and Moon, R. C.**, Inhibition of mammary and urinary bladder carcinogenesis by a retinoid and a maleic anhydride-devinyl ether copolymer (MVE-2), *Carcinogenesis*, 3, 1473, 1982.
38. **Lotan, R. and Nicolson, G. L.**, Inhibitory effects of retinoic acid or retinyl acetate on the growth of untransformed, transformed and tumor cells *in vitro*, *J. Natl. Cancer Inst.*, 59, 1717, 1977.
39. **Rudland, P. S., Paterson, F. C., Davies, A. C. T., and Warburton, M. J.**, Retinoid specific induction of differentiation and reduction of the DNA synthesis rate and tumor-forming ability of a stem cell line from a rat mammary tumor, *J. Natl. Cancer Inst.*, 70, 949, 1983.
40. **Maiorana, A. and Gullino, P. M.**, Effect of retinyl acetate on the incidence of mammary carcinomas and hepatomas in mice, *J. Natl. Cancer Inst.*, 64, 655, 1980.
41. **Welsch, C. W., DeHoog, J. V., and Moon, R. C.**, Inhibition of mammary tumorigenesis in nulliparous C3H mice by chronic feeding of the synthetic retinoid, N-(4-hydroxyphenyl)-retinamide, *Carcinogenesis*, 4, 1185, 1983.

42. **Welsch, C. W., Goodrich-Smith, M., Brown, C. K., and Crowe, N.,** Enhancement by retinyl acetate of hormone-induced mammary tumorigenesis in female GR/A mice, *J. Natl. Cancer Inst.,* 67, 935, 1981.

43. **Welsch, C. W. and DeHoog, J. V.,** Lack of an effect of dietary retinoids in chemical carcinogenesis of the mouse mammary gland: inverse relationship between mammary tumor cell anaplasia and retinoid effectiveness, *Carcinogenesis,* 5, 1301, 1984.

44. **Weiss, L. and Holyoke, E. C.,** Some effects of hypervitaminosis A on metastasis of spontaneous breast cancer in mice, *J. Natl. Cancer Inst.,* 43, 1045, 1969.

45. **Rettura, G., Schittek, A., Hardy, M., Levenson, S. M., Demetriou, A., and Seifter, E.,** Antitumor action of vitamin A in mice inoculated with adenocarcinoma cells, *J. Natl. Cancer Inst.,* 54, 1489, 1975.

46. **Rettura, G., Stratford, F., Levenson, S. M., and Seifter, E.,** Prophylactic and therapeutic actions of supplemental β-carotine in mice inoculated with C3HBA adenocarcinoma cells: lack of therapeutic action of supplemental ascorbic acid, *J. Natl. Cancer Inst.,* 69, 73, 1982.

47. **Telang, N. T., Banerjee, M. R., Iyer, A. P., and Kundu, A. B.,** Neoplastic transformation of epithelial cells in whole mammary gland in vitro, *Proc. Natl. Acad. Sci. U.S.A.,* 76, 5886, 1979.

48. **Dickens, M. S., Custer, R. P., and Sorof, S.,** Retinoid prevents mammary gland transformation by carcinogenic hydrocarbon in whole-organ culture, *Proc. Natl. Acad. Sci. U.S.A.,* 76, 5891, 1979.

49. **Dickens, M. S. and Sorof, S.,** Retinoid prevents transformation of cultured mammary glands by procarcinogens but not by many activated carcinogens, *Nature (London),* 285, 581, 1980.

50. **Chatterjee, M. and Banerjee, M. R.,** N-nitrosodiethylamine-induced nodule-like alveolar lesion and its prevention by a retinoid in Balb/c mouse mammary glands in the whole organ in culture, *Carcinogenesis,* 3, 801, 1982.

51. **Chatterjee, M. and Banerjee, M. R.,** Influence of hormones on N-(4-hydroxyphenyl)retinamide inhibition of 7,12-dimethylbenzanthracene transformation of mammary cells in organ culture, *Cancer Lett.,* 16, 239, 1982.

52. **Telang, N. T. and Sarkar, N. H.,** Long-term survival of adult mouse mammary glands in culture and their response to a retinoid, *Cancer Res.,* 43, 4891, 1983.

53. **Rieber, M. and Seyler, L.,** Growth inhibition of mouse mammary tumor cells by dexamethasone concurrent with enhanced endogenous protein phosphorylation: effects of retinoic acid, *Cell Biol. Int. Rep.,* 4, 1075, 1980.

54. **Mehta, R. G., Cerny, W. L., and Moon, R. C.,** Retinoids inhibit prolactin induced development of the mammary gland in vitro, *Carcinogenesis,* 4, 23, 1983.

55. **Lotan, R.,** Different susceptibilities of human melanoma and breast carcinoma cell lines to retinoic acid-induced growth inhibition, *Cancer Res.,* 39, 1014, 1979.

56. **Ueda, H., Takenawa, T., Millan, J. C., Gesell, M. S., and Brandes, D.,** The effects of retinoids on proliferative capacities and macromolecular synthesis in human breast cancer MCF-7 cells, *Cancer,* 46, 2203, 1980.

57. **Lacroix, A. and Lippman, M. E.,** Binding of retinoids to human breast cancer cell lines and their effects on cell growth, *J. Clin. Invest.,* 65, 586, 1980.

58. **Cassidy, J., Lippman, M., Lacroix, A., and Peck, G.,** Phase II trial of 13-cis-retinoic acid in metastatic breast cancer, *Eur. J. Cancer Clin. Oncol.,* 18, 925, 1982.

59. **Verma, A. K., Conrad, E. A., and Boutwell, R. K.,** Differential effects of retinoic acid and 7,8-benzoflavone on the induction of mouse skin tumors by the complete carcinogenesis process and by the initiation-promotion regimen, *Cancer Res.,* 42, 3519, 1982.

60. **Thompson, H. J., Becci, P. J., Brown, C. C., and Moon, R. C.,** Effect of the duration of retinyl acetate feeding on inhibition of 1-methyl-1-nitrosourea-induced mammary carcinogenesis in the rat, *Cancer Res.,* 39, 3977, 1979.

61. **Yuspa, S. H., Elgjo, K., Morse, M. A., and Wiebel, F. J.,** Retinyl acetate modulation of cell growth and carcinogen-cellular interactions, *Chem. Biol. Interact.,* 16, 251, 1977.

62. **Kohl, F. V. and Rudiger, H. W.,** In vitro untersuching an menschlichen zellkulturen zur tumorprophylaxe durch vitamin A, *J. Cancer Res. Clin. Oncol.,* 93, 149, 1979.

63. **Khandiya, K. L., Dogra, S. C., and Sharma, R. R.,** Body vitamin A status and binding of 7,12-dimethylbenzanthracene to rat mammary gland DNA, *Ind. J. Med. Res.,* 76, 702, 1982.

64. **Shoyab, M.,** Inhibition of the binding of 7,12-dimethylbenzanthracene to DNA of murine epidermal cells in culture by vitamin A and vitamin C, *Oncology,* 38, 187, 1981.

65. **Genta, V. M., Kaufman, D. G., Harris, C. C., Smith, J. M., Sporn, M. B., and Saffioti, U.,** Vitamin A deficiency enhances binding of benzopyrene to tracheal epithelial DNA, *Nature (London),* 247, 48, 1974.

66. **McCormick, A. M., Napoli, J. L., Yoshizawa, S., and DeLuca, H. F.,** 5,6-Epoxyretinoic acid as a physiological metabolite of retinoic acid in the rat, *Biochem. J.,* 186, 475, 1980.

67. **Levine, L. and Ohucki, K.,** Retinoids as well as tumor promoters enhance deacylation of cellular lipids and prostaglandin production in MDCK cells, *Nature (London),* 276, 274, 1978.

68. **Takenaga, K.,** Stimulation by retinoic acid of prostaglandin production and its inhibition by tumor promoters in mouse myeloid leukemia cells, *Gann,* 72, 488, 1981.

69. **Shoyab, M.,** Effect of prostaglandins and some anti-inflammatory drugs on the binding of 7,12-dimethylbenzanthracene to DNA of murine epidermal cells in culture, *Cancer Lett.*, 1, 155, 1979.

70. **Hill, D. L. and Shih, T. W.,** Vitamin A compounds and analogs as inhibitors of mixed-function oxidases that metabolize carcinogenic polycyclic hydrocarbons and other compounds, *Cancer Res.*, 34, 564, 1974.

71. **Miranda, C. L., Mukhtar, H., Bend, J. R., and Clihabra, R. S.,** Effects of vitamin A deficiency on hepatic and extrahepatic mixed-function oxidase and epoxide-metabolizing enzymes in guinea pig and rabbit, *Biochem. Pharmacol.*, 28, 2713, 1979.

72. **Narisawa, T., Reddy, B. S., Wong, C. G., and Weisburger, J. H.,** Effect of vitamin A deficiency on rat colon carcinogenesis by N-methyl-N-nitro-N-nitrosoguanidine, *Cancer Res.*, 36, 1379, 1976.

73. **Hauswirth, J. W. and Brizvela, B. S.,** The differential effects of chemical carcinogens on vitamin A status and on microsomal drug metabolism in normal and vitamin A deficient rats, *Cancer Res.*, 36, 1941, 1976.

74. **Kolamegham, R. and Krishnaswamy, K.,** Benzopyrene metabolism in vitamin A deficient rats, *Life Sci.*, 27, 33, 1980.

75. **Welsch, C. W., Goodrich-Smith, M., Brown, C. K., and Roth, L.,** The prophylaxis of rat and mouse mammary gland tumorigenesis by suppression of prolactin secretion: a reappraisal, *Breast Cancer Res. Treat.*, 1, 225, 1981.

76. **Welsch, C. W., Goodrich-Smith, M., Brown, C. K., Mackie, D., and Johnson, D.,** 2-Bromo-α-ergocryptine (CB-154) and tamoxifen (ICI 46,474) induced suppression of the genesis of mammary carcinomas in female rats treated with 7,12-dimethylbenzanthracene (DMBA): a comparison, *Oncology*, 39, 88, 1982.

77. **Mehta, R. G. and Moon, R. C.,** Inhibition of DNA synthesis by retinyl acetate during chemically induced mammary carcinogenesis, *Cancer Res.*, 40, 1109, 1980.

78. **Sankaran, L. and Topper, Y. J.,** Effect of vitamin A deprivation on maintenance of rat mammary tissue and on the potential of the epithelium for hormone-dependent milk protein synthesis, *Endocrinology*, 111, 1061, 1982.

79. **Bashor, M. M., Toft, D. O., and Chytil, F.,** In vitro binding of retinol to rat tissue compounds, *Proc. Natl. Acad. Sci. U.S.A.*, 70, 3433, 1973.

80. **Sani, B. B. and Corbett, T. H.,** Retinoic acid binding protein in normal tissue and experimental tumors, *Cancer Res.*, 37, 209, 1977.

81. **Ong, D. E., Page, D. L., and Chytil, F.,** Retinoic acid binding protein: occurrence in human tumors, *Science*, 190, 60, 1975.

82. **Sani, B. P.,** Localization of retinoic acid binding protein in nuclei, *Biochem. Biophys. Res. Commun.*, 75, 7, 1977.

83. **Takase, S., Ong, D. E., and Chytil, F.,** Cellular retinol-binding protein allows specific interaction of retinol with the nucleus *in vitro*, *Proc. Natl. Acad. Sci. U.S.A.*, 76, 2204, 1979.

84. **Mehta, R. G., Cerny, W. L., and Moon, R. C.,** Nuclear interactions of retinoic acid-binding protein in chemically induced mammary adenocarcinoma, *Biochem. J.*, 208, 731, 1982.

85. **Huber, P. R., Geyer, E., Kung, W., Matter, A., Torhorst, J., and Eppenberger, U.,** Retinoic acid binding protein in human breast cancer and dysplasia, *J. Natl. Cancer Inst.*, 61, 1375, 1978.

86. **Kung, W. M., Geyer, E., Uppenberger, U., and Huber, P. R.,** Quantitative estimation of cellular retinoic acid binding protein activity in normal, dysplastic, and neoplastic human breast tissue, *Cancer Res.*, 40, 4265, 1980.

87. **Takenawa, T., Ueda, H., Brandes, D., and Millan, J. C.,** Retinoic acid binding protein in a human cell (MCF-7) from breast carcinoma, *Lab. Invest.*, 42, 490, 1980.

88. **Mehta, R. G., Kute, T. E., Hopkins, M., and Moon, R. C.,** Retinoic acid binding proteins and steroid receptor levels in human breast cancer, *Eur. J. Cancer Clin. Oncol.*, 18, 221, 1982.

89. **Ong, D. E. and Chytil, F.,** Presence of cellular retinol and retinoic acid binding proteins in experimental tumors, *Cancer Lett.*, 2, 25, 1976.

90. **Sani, B. P. and Corbett, T. H.,** Retinoic acid binding protein in normal tissues and experimental tumors, *Cancer Res.*, 37, 209, 1977.

91. **Sani, B. P. and Titus, B. C.,** Retinoic acid-binding protein in experimental tumors and in tissues with metastatic tumor foci, *Cancer Res.*, 37, 4031, 1977.

92. **Mehta, R. G., Cerny, W. L., and Moon, R. C.,** Distribution of retinoic acid-binding proteins in normal and neoplastic mammary tissues, *Cancer Res.*, 40, 47, 1980.

93. **Mehta, R. G., McCormick, D. L., Cerny, W. L., and Moon, R. C.,** Correlation between retinoid inhibition of N-methyl-N-nitrosourea-induced mammary carcinogenesis and levels of retinoic acid binding proteins, *Carcinogenesis*, 3, 89, 1982.

94. **Moon, R. C. and Mehta, R. G.,** Retinoid binding in normal and neoplastic mammary tissue, in *Hormones and Cancer*, Leavitt, W. W., Ed., Plenum Press, New York, 1982, 231.

95. **Bunk, M. J., Telang, N. T., and Sarkar, N. H.,** Effect of malignant transformation upon the cellular retinoid binding proteins in cultured murine mammary cells, *Cancer Lett.*, 20, 83, 1983.

96. **Lotan, R., Ong, D. E. and Chytil, F.,** Comparison of the level of cellular retinoid-binding proteins and susceptibility to retinoid-induced growth inhibition of various neoplastic cell lines, *J. Natl. Cancer Inst.,* 64, 1259, 1980.

97. **Libby, P. R. and Bertram, J. S.,** Lack of intracellular retinoid-binding proteins in a retinol-sensitive cell line, *Carcinogenesis,* 3, 481, 1982.

98. **Ong, D. E. and Chytil, F.,** Changes in levels of cellular retinol- and retinoic acid-binding proteins in liver and lung during perinatal development of the rat, *Proc. Natl. Acad. Sci. U.S.A.,* 73, 3976, 1976.

99. **Kylberg, H. K., Ong, D. E., and Chytil, F.,** Cellular retinol binding protein during postnatal development of the rat small intestine, *Biol. Neonate,* 39, 100, 1981.

100. **Fell, H. B. and Dingle, J. T.,** Studies on the mode of action of excess vitamin A. VI. Lysosomal protease and the degradation of cartilage matrix, *Biochem. J.,* 87, 403, 1967.

101. **Mayer, H., Bollag, W., Hanni, R., and Ruegg, R.,** Retinoids: a new class of compounds with prophylactic and therapeutic activities in oncology and dermatology, *Experientia,* 34, 1105, 1978.

102. **Shamberger, R. J.,** Inhibitory effect of vitamin A on carcinogenesis, *J. Natl. Cancer Inst.,* 47, 667, 1971.

103. **DeLuca, L. M.,** The direct involvement of vitamin A in glycosyl transfer reactions in mammalian membranes, *Vitam. Horm.,* 35, 1, 1977.

104. **Warren, L., Buck, C. A., and Tuszynski, G. P.,** Glycopeptide changes and malignant transformation. A possible role for carbohydrate in malignant behavior, *Biochim. Biophys. Acta,* 516, 97, 1978.

105. **Bertram, J. S., Mordan, L. J., Domanska-Janik, K., and Bernacki, R. J.,** Inhibition of in vitro neoplastic transformation by retinoids, in *Molecular Interactions of Nutrition and Cancer,* Arnott, M. S., Van Eys, J., and Wang, Y. M., Eds., Raven Press, New York, 1982, 315.

106. **Adams, S., Sasak, W., Dick, L. D., and DeLuca, L. M.,** Studies on the mechanism of retinoid-induced adhesion of spontaneously transformed mouse fibroblasts, *Acta Vitaminol. Enzymol.,* 5, 3, 1983.

107. **Mikherjee, B. B., Mobry, P. M., and Pena, S. D. J.,** Retinoic acid induces anchorage- and density dependent growth without restoring normal cytoskeleton, EGF binding, fibronectin content and ODC activity in a retrovirus-transformed mouse cell line, *Exp. Cell Res.,* 138, 95, 1982.

108. **Chertow, B. S., Baranestsky, N. G., Sivitz, W. I., Mida, P., Webb, M. D., and Shik, J. C.,** Cellular mechanisms of insulin release. Effects of retinoids on rat islet cell-to-cell adhesion, reaggregation and insulin release, *Diabetes,* 32, 568, 1983.

109. **Levin, L. V., Clark, J. N., Ouill, H. R., Newberne, P. M., and Wolf, G.,** Effect of retinoic acid on the synthesis of glycoproteins of mouse skin tumors during progression from promoted skin through papillomas to carcinomas, *Cancer Res.,* 43, 1724, 1983.

110. **Mordan, L. J. and Bertram, J. S.,** Retinoid effects on cell-cell interactions and growth characteristics of normal and carcinogen-treated C3H/10T$^1/_2$ cells, *Cancer Res.,* 43, 567, 1983.

111. **Elias, P. M., Grayson, S., Caldwell, T. M., and McNutt, N. S.,** Gap junction proliferation in retinoic acid treated human basal cell carcinoma, *Lab. Invest.,* 42, 469, 1980.

112. **Shuin, T., Nishimura, R., Noda, K., Umeda, M., and Ono, T.,** Concentration-dependent differential effect of retinoic acid on intercellular metabolic cooperation, *Gann,* 74, 100, 1983.

113. **Zile, M. H. and Cullum, M. E.,** The function of vitamin A: current concepts, *Proc. Soc. Exp. Biol. Med.,* 172, 139, 1983.

114. **Todaro, G. J., DeLarco, J. E., and Sporn, M. B.,** Retinoids block phenotypic cell transformation produced by sarcoma growth factor, *Nature (London),* 276, 272, 1978.

115. **Welsch, C. W. and Nagasawa, H.,** Prolactin and murine mammary tumorigenesis: a review, *Cancer Res.,* 37, 951, 1977.

116. **Welsch, C. W., Iturri, C., and Brennan, M. J.,** DNA synthesis of human, mouse and rat mammary carcinomas in vitro. Influence of insulin and prolactin, *Cancer,* 38, 1272, 1976.

117. **Harper, R. A. and Savage, C. R.,** Vitamin A potentiates the mitogenic effect of epidermal growth factor in cultures of normal adult human skin fibroblasts, *Endocrinology,* 107, 2113, 1980.

118. **Dicker, P. and Rozengurt, E.,** Retinoids enhance mitogenesis by tumour promoter and polypeptide growth factors, *Biochim. Biophys. Res. Commun.,* 91, 1203, 1979.

119. **Speers, W. C.,** Conversion of malignant murine embryonal carcinomas to benign teratomas by chemical induction of differentiation in vivo, *Cancer Res.,* 42, 1843, 1982.

120. **Rillema, J. A.,** Mechanism of prolactin action, *Fed. Proc.,* 39, 2593, 1980.

121. **Cohen, L. A. and Chan, P. C.,** Intracellular cAMP levels in normal rat mammary gland and adenocarcinoma. *In vivo* vs. *in vitro, Life Sci.,* 16, 107, 1975.

122. **Strickland, S., Smith, K. K., and Marotti, K. R.,** Hormonal induction of differentiation in teratocarcinoma stem cells: generation of parietal endoderm by retinoic acid and dibutyryl cAMP, *Cell,* 21, 347, 1980.

123. **Strickland, S. and Mahdavi, V.,** The induction of differentiation in teratocarcinoma stem cells by retinoic acid, *Cell,* 15, 393, 1978.

124. **Olsson, I. L., Breitman, T. R., and Gallo, R. C.,** Priming of human myeloid leukemic cell lines HL-60 and U-937 with retinoic acid for differentiation: effects of cyclic adenosine 3',5'-monophosphate-inducing agents and a T-lymphocyte-derived differentiation factor, *Cancer Res.,* 42, 3923, 1982.

125. **Plet, A., Evain, D., and Anderson, W. B.**, Effect of retinoic acid treatment of F9 embryonal carcinoma cells on the activity and distribution of cyclic AMP-dependent protein kinase, *J. Biol. Chem.*, 257, 889, 1982.

126. **Ludwig, K. W., Lowey, B., and Niles, R. M.**, Retinoic acid increases cyclic AMP-dependent protein kinase activity in murine melanoma cells, *J. Biol. Chem.*, 255, 5999, 1980.

127. **Fontana, J. A., Emler, C. A., McClung, J. K., Butcher, F. R., and Durham, J. P.**, Modulation of cAMP-dependent protein kinases (cAMP-dPK) during retinoic acid (RA) and 12-O-tetradecanoyl-phorbol-13 acetate (TP) induced differentiation of HL60 cells, *Proc. Am. Assoc. Cancer Res.*, 12, 12, 1983.

128. **Welsch, C. W. and DeHoog, J. V.**, Influence of cholera toxin on the growth and development of N-methyl-N-nitrosourea-induced rat mammary carcinomas, *Cancer Lett.*, 20, 61, 1983.

129. **Westin, E. H., Wong-Staal, F., Gelmann, E. P., DallaFavera, R., Papas, T. S., Lautenberger, J. A., Eva, A., Reddy, E. P., Tronick, S. R., Aaronson, S. A., and Gallo, R. C.**, Expression of cellular homologues of retroviral onc genes in human hematopoietic cells, *Proc. Natl. Acad. Sci.*, 79, 2490, 1982.

130. **Sidell, N.**, Retinoic acid-induced growth inhibition and morphologic differentiation of human neuroblastoma cells *in vitro*, *J. Natl. Cancer Inst.*, 68, 589, 1981.

131. **Dennert, G., Crowley, C., Kouba, J., and Lotan, R.**, Retinoic acid stimulation of the induction of mouse killer T-cells in allogeneic and syngeneic systems, *J. Natl. Cancer Inst.*, 62, 89, 1979.

132. **Floersheim, G. L. and Bollag, W.**, Accelerated rejection of skin homografts by vitamin A acid, *Transplantation*, 14, 564, 1972.

133. **Jurin, M. and Tannock, I. F.**, Influence of vitamin A on immunological response, *Immunology*, 23, 283, 1972.

134. **Dresser, D. W.**, Adjuvanticity of vitamin A, *Nature (London)*, 217, 527, 1968.

135. **Medawar, P. B. and Hunt, R.**, Anti-cancer action of retinoids, *Immunology*, 42, 349, 1981.

136. **Athanassiades, T. J.**, Adjuvant effect of vitamin A palmitate and analogs on cell-mediated immunity, *J. Natl. Cancer Inst.*, 67, 1153, 1981.

137. **Rhodes, J. and Oliver, S.**, Retinoids as regulators of macrophage function, *Immunology*, 40, 467, 1980.

138. **Meltzer, M. S. and Cohen, B. E.**, Tumor suppression by mycobacterium bovis (strain BCG) enhanced by vitamin A, *J. Natl. Cancer Inst.*, 53, 585, 1974.

139. **Tannock, I. F., Suit, H. O., and Marshall, N.**, Vitamin A and the radiation response of experimental tumours: an immune mediated effect, *J. Natl. Cancer Inst.*, 48, 731, 1972.

Chapter 2

SELENIUM AND MURINE MAMMARY TUMORIGENESIS

Daniel Medina

IN MEMORIAM

This review is dedicated to the memory of Dr. A. Clark Griffin, who died unexpectedly on December 13, 1982. Dr. Griffin spent most of his career in cancer research and during the latter years, he was a leader in the area of mechanisms of action of selenium-mediated inhibition of tumorigenesis. He is remembered not only as a friend, but also as a gentle man who enthusiastically encouraged the pursuit of high quality science. I am indebted to Dr. Griffin for encouraging me to continue examining the relationship between selenium and mammary tumorigenesis.

TABLE OF CONTENTS

I. INTRODUCTION

The chemical element, selenium (Se), was discovered by Jon Berzelius in 1818. Table 1 lists some essential facts about the chemistry, distribution, and uses of selenium. Selenium lies in Group VIA in the periodic table of elements between sulfur and tellurium. The chemical and physical characteristics of selenium and sulfur are similar; however, despite these similarities, selenium compounds tend to be reduced, whereas sulfur compounds tend to be oxidized in living systems.[1] In nature, specific plants incorporate selenium from the soil and accumulate selenium as both the organic and inorganic forms.[2] The distribution of selenium in the soil varies geographically with low selenium areas roughly east of the Mississippi and north of the Ohio rivers, as well as certain areas along the southeast U.S. Atlantic coastline and northwest U.S. Pacific coastline. In certain areas, plants (grasses, grains) accumulate enough selenium so they are toxic to grazing livestock.[3] An excellent review of the chemistry and biochemistry of selenium can be found in Shamberger's book on selenium.[1]

The importance of selenium as an essential nutrient has been recognized in the field of animal husbandry for a considerable length of time. Both selenium deficiency and selenium toxicity syndromes in livestock have been extensively documented.[1-4] Selenium deficiency manifests itself as liver necrosis (rats), exudative diathesis (chickens), pancreatic necrosis (chickens), muscular dystrophy (sheep, cattle, horses), and infertility (ewes, rats).[1,5-7] Selenium toxicity manifests itself as blind staggers, a disease characterized by CNS disorders, and alkali disease, a disease which exhibits symptoms of retarded growth, hair loss, weight loss, deformed hoofs, and death.[4,8]

The biochemical basis for the essentiality of selenium was resolved in 1973 when Rotruck et al.[9] suggested that the enzyme glutathione peroxidase (GSH-PX) (EC 1.11.19) was a seleno-enzyme. Additional experiments showed there were four atoms of selenium per molecule of GSH-Px.[10] The enzyme is present in all tissues examined in both avian and mammalian species, including erythrocyte, liver, kidney, heart, lung, stomach, spleen, brain, and mammary gland. The enzymes found in the mammary gland, liver, and erythrocyte have the same kinetics and are similar in molecular weights.[1,11] GSH-Px catalyzes the following reaction:[9]

$$ROOH + 2GSH \xrightarrow{\text{GSH-Px}} ROH + GSSG$$

The substrate (ROOH) can be either hydrogen peroxide or lipid peroxide. Thus, the enzyme is important in preventing peroxidative damage to cellular macromolecules. Although other proteins have been tentatively identified as selenoproteins in eukaryotic cells, GSH-Px is the only protein functionally characterized.

The use of selenium as a cancer chemopreventive agent is a relatively new idea. Selenium supplementation to the diet inhibits chemical carcinogen-induced tumors in skin, liver, colon, and mammary gland and also MMTV-induced tumors in the mammary gland.[12-42] Table 2 summarizes the results of experiments which have examined the effect of selenium on tumorigenesis in various rodent model systems.

Of the 36 experiments, 30 have been published since 1976; 32 demonstrated an inhibitory effect of selenium supplementation, 2 showed a slight enhancing effect, and 2 showed no effect of selenium. A comment on the initial experiment reported by Nelson et al.[12] is required at this point. This experiment documented an increase in liver adenomas in rats fed a diet enriched in seleniferous corn and wheat or inorganic selenide. Concurrent with the adenomas was an extensive amount of liver cirrhosis. This experiment was repeated twice in very large-scale carefully controlled experiments in 1949 by Clayton and Baumann[13] and in 1972 by Harr et al.[14] Both groups of investigators failed to find any evidence for the

Table 1
SELECTED CHARACTERISTICS OF SELENIUM

Chemistry:	molecular weight = 79, isotope = 75 (gamma emitter, 121-day half-life)
Nature:	Earth's crust, 17th most abundant element
Diet:	grain (cereals), seafood, poultry, legumes, organ meats; average U.S. intake — 100 μg/day
Industrial uses:	xerography, resistors, glass tint (red)
Deficiency states:	pancreatic/liver necrosis, muscular dystrophy, cardiomyopathy
Toxicity:	blind staggers, alkali disease

enhancement of carcinogenicity by selenium; on the contrary, they provided the first evidence that selenium inhibited chemical carcinogenesis in the liver. As will be discussed in greater detail, selenium inhibits both the initiation and postinitiation phases of chemical carcinogen. Thus, the evidence is impressive that selenium supplementation to the diet exerts a strong inhibitory effect on the tumorigenic process.

The epidemiological evidence for the identification of selenium as a possible protective factor in human tumorigenesis is scarce but suggestive. The data is based on correlative studies carried out by several investigators. Shamberger and co-workers[43] have compared the age-specific mortality rates for 55- to 64-years-old interval with the published selenium concentration of forage crops in the U.S. There was an inverse correlation between mortality rate and selenium concentration in forage crops for GI, bladder, lung, breast, and uterine cancer. In another study, the same group compared the mortality rates for 14 specific cancers in man who lived in 17 paired cities located in low or high selenium areas where a variety of epidemiological factors were similar (i.e., urbanization, air pollution, racial heterogeneity of the population).[44] The mortality rates for organs of the GI tract as well as bladder and kidney were higher in low selenium areas.

Schrauzer and co-workers[45] correlated selenium per capita intake from food consumption data with mortality rate and found inverse correlations between selenium intake and mortality for cancers of the intestine, rectum, breast, prostate, lung, ovary, and leukemia. Jansson and Jacobs[46] reported that Seneca County in New York had a lower incidence and mortality rate of colorectal cancer than other counties in New York. The selenium level in the water system was twice the average level for New York.

Several investigators have reported that the selenium levels in whole blood of cancer patients (colon, breast, pancreas, head, and neck) was significantly lower compared to the blood of normal or noncancer diseased patients.[47-50] In one study of head and neck patients, the selenium levels were decreased in erythrocytes and were increased in plasma.[50]

Although the above cited data (for a thorough review, see Reference 1) are suggestive, further epidemiological studies are needed to document the relationship between cancer mortality and selenium intake. In particular, case-control studies are needed based on appropriately stratified populations and exhaustive dietary histories. Additionally, more information is needed on the levels and bioavailability of selenium in different foods.

The remaining part of this review will concentrate on the effects of selenium on mammary gland tumorigenesis. In general, the results seen in the mammary gland system have also been seen in other systems when similar questions have been examined.

II. INHIBITION OF MURINE MAMMARY TUMORIGENESIS

A. Inhibition of Tumorigenesis

The inhibition of mammary tumorigenesis is manifested as an inhibition of mammary adenocarcinoma formation. Inhibition is generally expressed as the percent tumor incidence (i.e., percent of mice/rats with tumors) or total number of tumors. The former parameter is

Table 2
EFFECT OF SELENIUM SUPPLEMENTATION ON TUMORIGENESIS

Species	Organ	Carcinogen[a]	Route of administration of selenium	Effect on tumorigenesis[b]	Ref.
Rat	Liver	None	Diet	↑ (20)[c]	12
Rat	Liver	DAB	Diet	↓ (50)	13
Rat	Liver	AAF	Diet	↓ (50)	14
Rat	Liver	DAB	Water	↓ (50)	15
Rat	Liver	AAF	Water	↓ (59)	16
Rat	Liver	DAB	Water	↓ (52—82)	17
Rat	Colon	DMH	Water	↓ (72)	18
Rat	Colon	MAM	Water	↓ (43)	18
Rat	Colon	AOM	Water	↓ (52)	19
Rat	Colon	DMH	Water	↓ (38)	20
Rat	Colon	BOP	Diet	↓ (43)	21
Rat	Colon/Cecum	DMH	Water	↓ (NA)	22
Rat	Colon	DMH	Water	↓ (77)	23
Rat	Mammary gland	AAF	Diet	↓ (70)	14
Rat	Mammary gland	MNU	Diet	↓ (29—42)	24
Rat	Mammary gland	DMBA	Water	↓ (53)	25
Rat	Mammary gland	MNU	Diet	↓ (23)	26
Rat	Mammary gland	DMBA	Diet	↓ (62—79)	27
Rat	Mammary gland	DMBA	Diet	↓ (42—66)	28
Rat	Mammary gland	DMBA	Diet	↓ (49)	29
Rat	Mammary gland	DMBA	Diet	↓ (23—48)	30
Rat	Mammary gland	Ad-9	Water	↑ (67)[d]	22
Rat	Mammary gland	DMBA	Diet	↓ (17)	31
Mouse	Mammary gland	MMTV-S	Water	↓ (88)	32
Mouse	Mammary gland	MMTV-S	Water	↓ (56)	33
Mouse	Mammary gland	MMTV-S	Diet, water	↓ (49—88)	34
Mouse	Mammary gland	MMTV-S	Diet	↓ (65)	35
Mouse	Mammary gland	MMTV-S	Water	↓ (42—85)	36
Mouse	Mammary gland	MMTV-P	Water	No effect	25
Mouse	Mammary gland	DMBA	Water	↓ (42—85)	37
Mouse	Mammary gland	DMBA	Water	↓ (62—91)	38
Mouse	Mammary gland	DMBA	Diet	↓ (71)	39
Mouse	Mammary gland	MMTV	Diet	↓ (82)	39
Mouse	Mammary gland	DMBA	Water	↓ (40—81)	40
Mouse	Skin	DMBA	Ectopically, diet	↓ (36—50)	41
Hamster	Trachea	MNU	Diet	No effect	42

[a]   DAB, 3-methyl-4-dimethylaminoazobenzene; AAF, 2-acetylaminofluorine; DMH, 1,2-dimethylhydrazine; MAM, methylazoxymethanol acetate; AOM, azoxymethane; BOP, bis(2-oxopropyl)-nitrosamine; MNU, methylnitrosourea; DMBA, 7,12-dimethylbenzanthracene; Ad-9, adenovirus type 9; MMTV, mouse mammary tumor virus.

[b]   ↓, decrease; ↑, increase. The numbers in parentheses indicate the percentage inhibition or increase of tumorigenesis compared to control animals. NA, = not available.

[c]   Increase in cirrhosis and adenomas.

[d]   ↑, in fibroadenomas.

used in experiments with mice since multiple tumors are rare; thus interpretation of the data is straight-forward. Both parameters are used in experiments with rats; thus one parameter (usually the total number of tumors) may change more significantly than the other. Thus, interpretation of the data can be complicated, and in this review, inhibition was considered significant if only one parameter changed (see Chapter 6, Volume I for further discussion of this question.) For instance, in one experiment on the effect of selenium administered during the postinitiation stage of carcinogenesis, the mammary cancer incidence was decreased by 10% but the total number of tumors decreased by 42%. The authors interpreted the 10% reduction in mammary cancer incidence as probably contributed by reduced food intake, since they demonstrated that food restriction leading to an 8% loss in body weight caused a 10 to 15% loss in rate of tumor occurrence.[24] A similar finding was also reported by Ip.[27] However, the majority of the 42% decrease in total number of tumors was attributed to an independent inhibitory effect of selenium.

Selenium-mediated inhibition of mammary tumorigenesis has been demonstrated for chemical carcinogen-induced tumors in the rat (DMBA, MNU, AAF) and mouse (DMBA) and for MMTV-induced tumors in several inbred mouse strains (C3H/St, BALB/cfC3H, BALB/cV) (see Table 2). The amount of selenium required for inhibition of tumorigenesis has been examined in several dietary experiments in which the variables of caloric intake, weight gain, and total nutrient intake were carefully controlled. Depending on the carcinogen, a dose of 0.5 to 2.5 ppm Se reduced the tumor burden by at least 50% in both rats and mice. In rats, 0.5 ppm Se added to the diet inhibited AAF-induced tumorigenesis in the mammary gland and liver,[14] whereas 2.5 ppm Se was necessary to inhibit DMBA-induced mammary tumorigenesis.[27] In mice, 0.5 ppm dietary Se inhibited DMBA-induced tumorigenesis in BALB/c mice[39] and MMTV-induced tumorigenesis in C3H/St mice.[34]

Selenium inhibition can be overcome by the strength of the carcinogen insult and/or the type of enhancing stimulus. (In this review, the terms "initiation-postinitiation" are used instead of "initiation-promotion", since the latter concept is difficult to demonstrate conclusively with the agents used in the mammary tumor system. Similarly, the term "enhancement" rather than the term "promotion" is used through the text) (see Chapter 6 Volume I for further discussion). For instance, 2.5 ppm dietary Se inhibited equally well the tumorigenic response induced by 5 or 10 mg DMBA if the rats were fed a low-fat (5% corn oil) diet; however, a greater amount of dietary Se (5.0 ppm) was needed to inhibit the tumorigenic response induced by 10 mg (but not 5 mg DMBA) if the rats were fed a high-fat diet (25% corn oil). Thus, in this experiment, the tumorigenic process occurring under the influence of a high concentration of DMBA and polyunsaturated fats required a greater amount of selenium to achieve a significant inhibition.[27] In contrast, Thompson et al.[30] reported that 2 ppm dietary Se given in a Torula yeast diet inhibited the tumorigenic response induced by a high dose of DMBA (15 mg) but not the response induced by a low dose of DMBA (7.5 mg). An explanation for this apparent contradiction in the latter study lies in the suggestion that the relationship between the dose of carcinogen and the tumorigenic response is nonlinear, perhaps sigmoidal.[30] Thus, it becomes important to design and compare experiments where the carcinogen-induced response is operating within the linear portion of the response curve. It was speculated by Thompson et al.[30] that their low dose (7.5 mg) was in the nonlinear portion of their tumorigenic response curve. In DMBA-treated mice, selenium (6 ppm in the water) inhibited the tumorigenic response to a greater degree in mice given 2 mg DMBA as compared to 6 mg DMBA for both (C57BLxDBA2/f)F_1 and BALB/c mice.[37]

Mouse mammary tumorigenesis is also influenced greatly by genetics and by hormone status of the host. These influences also appeared in the effects of selenium. For instance, whereas MMTV-positive mouse strains of C3H/St, BALB/cfC3H, and BALB/cV were responsive to selenium-mediated tumorigenesis,[32-36,39] the MMTV-positive GR/A strain was not responsive to selenium-mediated tumorigenesis.[25] Similarly selenium supplementation inhibited DMBA-induced mammary tumorigenesis in virgin $BD2F_1$ mice by 61% but by only 23% in hormone-stimulated $BD2F_1$ mice.[37] It is thoroughly documented in the scientific literature that the mammary tumor response depends not only on species of animal, but the strain of mice (i.e., C3H vs. BALB/c) and even the substrain of rats (i.e., Sprague-Dawley [S-D] Charles-River vs. S-D Taconic Farms), the type of diet used (casein, Torula yeast, chow), and the specific carcinogen (DMBA, MNU, AAF). These results indicate that hard and firm generalizations cannot be derived regarding the quantitative aspects of selenium-mediated inhibition of tumorigenesis, a conclusion which is not surprising.

In the majority of experiments on selenium-mediated inhibition of tumorigenesis, both mammary and nonmammary, the selenium supplementation had been administered throughout the course of the experiment. Recent experiments have demonstrated that selenium prevents and delays the tumorigenic response in the postinitiation phase only as long as it is provided as a supplement. Second, selenium can inhibit either the initiation and postinitiation independently. In support of the first conclusion, the experiments by Schrauzer et al.[35] in C3H/St mice indicated that the mammary tumor incidence in mice switched at 13.8 months from a diet containing 0.15 ppm Se to one containing 1.0 ppm Se decreased from 77 to 46%. Conversely, the mammary tumor incidence in mice switched from a diet containing 1.0 ppm to one containing 0.15 ppm Se increased from 27 to 69%.

In support of the second conclusion, Thompson and co-workers[24,26,30] in separate experiments, demonstrated that dietary selenium supplementation administered just around DMBA feeding (-3 to $+2$ weeks) or administered starting 7 days after MNU feeding significantly inhibited mammary tumorigenesis in the rat. A similar finding was reported by Welsch et al.,[25] who administered selenium in the water from 30 to 90 days after DMBA treatment. Since these experiments were done at different times and different carcinogens, it is difficult to compare the relative effectiveness of selenium as an inhibitor of initiation vis-a-vis postinitiation events. This latter question has been addressed in part in experiments by Ip in the rat[28] and Medina and Lane in the mouse.[40] In the rat experiment, the efficacy of selenium was compared when given at times before ($-$) and/or after ($+$) DMBA administration. Dietary selenium was administered at the following weekly time periods: -2 to $+24$; -2 to $+2$; $+2$ to $+24$; $+2$ to $+12$; $+12$ to $+24$; and -2 to $+12$. All rats were killed at 24 weeks after DMBA administration. Several conclusions can be drawn from these results: (1) selenium inhibits both the initiation and postinitiation stages of tumorigenesis. Thus, 5 ppm dietary Se given at -2 to $+2$ weeks inhibited the total number of tumors by 41% as compared to 66% when given from -2 to $+24$ weeks. Dietary selenium given at $+2$ to $+24$ weeks inhibited the total number of tumors by 47%. The extent of inhibition when selenium was given around the time of initiation was remarkable since selenium was given for such a short period of time. There was no indication that the inhibition was reversible since the slope of the line illustrating tumor occurrence over time was linear rather than biphasic. (2) The inhibitory effect of selenium during postinitiation events ($+2$ to $+12$) was reversible since the slope of the line illustrating tumor occurrence over time was biphasic. As originally shown by Schrauzer et al.,[35] the rate of tumor appearance accelerated once selenium was withdrawn. Conversely, if selenium was added late ($+12$ to $+24$), the rate of tumor occurrence over time decreased. Maximal inhibition of tumorigenesis required the continuous presence of selenium.

A similar experiment was done with mice. The mammary tumor incidence in mice exposed to selenium in the postinitiation phase of tumorigenesis (14 weeks of age and thereafter) was 8% which was comparable to the incidence (13%) in mice exposed to selenium for the total tumorigenic process (6 weeks of age and thereafter). The mammary tumor incidence in mice exposed to selenium between weeks 6 and 14, the period encompassing carcinogen treatment, was 25%. In the absence of selenium, the mammary tumor incidence was 42%. The results of both experiments demonstrate that selenium is an effective chemopreventive agent in both the initiation and postinitiation phases of chemical carcinogenesis. Similar conclusions can be drawn from experiments on skin,[41] liver,[17] and colon,[20] where selenium has been shown to inhibit either of the two stages of carcinogenesis. These latter experiments were incomplete since they were designed to test where selenium inhibited either initiation or postinitiation events, but not both events separately.

Selenium deficiency, as well as selenium supplementation, can influence DMBA-induced mammary tumorigenesis in the rat.[51] In rats maintained on a polyunsaturated fat diet (5 or 25% corn oil) fed a low dose of DMBA (5 mg), selenium deficiency enhanced mammary tumorigenesis (tumor yield) by 85 to 158%, respectively. In rats fed a high dose of DMBA (10 or 15 mg) on a low polyunsaturated fat diet (1% corn oil, 24% saturated fat), selenium deficiency enhanced mammary tumorigenesis (tumor yield) by 58 to 84%, respectively. Selenium deficiency did not influence mammary tumorigenesis in rats maintained on a very low polyunsaturated fat diet (1%) and fed a low dose of DMBA (5 mg). These results suggested that the antioxidant property of selenium might be a possible mechanism by which selenium exerts a protective effect.

The interaction of selenium and other nutrients/trace elements on mammary tumorigenesis has been investigated in two different model systems. The synergistic effects of selenium and vitamin A (retinyl acetate) and E were examined in carcinogen-treated rats.[26,29,31] Rats treated with DMBA and fed selenium (4 ppm Se) and retinyl acetate (250 mg/kg during the entire tumorigenic process) had a tumor yield 8% of control rats as compared to 51 and 36% for rats fed either selenium or retinyl acetate alone. Interestingly, once the selenium-retinyl acetate supplement was withdrawn from the diet, tumors appeared at the same rate as in control rats, providing further support to the idea that selenium supplementation has to be continuously present to be effective.[29] A second study reported similar results. Rats treated with MNU and fed selenium (4 ppm) and retinyl acetate (300 mg/kg) during only the postinitiation stage had a tumor yield 47% of control rats compared to 77 and 66% for the rats fed selenium or retinyl acetate alone.[26]

In the above-mentioned experiments, retinyl acetate had an independent and potent inhibitory effect on mammary tumorigenesis. Vitamin E did not exhibit an independent inhibitory effect on DMBA-induced mammary tumorigenesis. However, rats treated with DMBA and fed selenium (2.5 ppm) and vitamin E (1000 mg/kg) throughout the carcinogenic process had a tumor yield of 45% compared to 68 and 92% for rats fed selenium and vitamin E alone. Of interest was the result that vitamin E exerted its synergistic effect with selenium if given during only the postinitiation stage but not if given during only the initiation stage of carcinogenesis.[31] These results indicate that the chemopreventive or anticarcinogenic efficacy of low levels of several agents is a rational approach for dietary control of mammary carcinogenesis. The use of low levels of several agents provides a true synergistic effect and avoids the possible deleterious effects associated with very high doses of nutrients and trace elements.

The interaction of selenium and zinc in mouse mammary tumorigenesis was examined in a brief study.[33] Zinc appeared to behave as a physiological antagonist to selenium; it completely prevented the selenium-mediated inhibition of mammary tumorigenesis. Unfortunately, mice maintained on a zinc supplement were not run as controls; therefore, it could not be determined whether zinc exerted an independent tumor-enhancing effect in this model system. However, the results did emphasize the importance of trace element interactions in tumorigenesis. It is well established that cadmium, mercury, lead, and zinc all modulate the effects of selenium and a careful study on these interactions with respect to carcinogenesis would be worthwhile. In this respect, a retrospective study did note that there was a direct correlation between breast cancer mortality and apparent dietary zinc intake from data of 28 countries.[33]

B. Stage Specificity

Mouse mammary tumorigenesis is characterized by the presence of discrete foci of alveolar and ductal hyperplasias which are precursor lesions to mammary tumors.[52,53] Thus, the tumorigenic process can be divided into several stages which are illustrated in the following scheme.

Normal → I-Preneoplasia → Preneoplasia → Neoplasia →

The biological characteristics of these different stages have been reviewed extensively.[52-55] From this scheme, it can be seen that selenium theoretically can exert an effect at one of several different steps in the transition from normal to neoplasia. Initially, selenium could prevent the interaction of a carcinogen with a cell, thereby preventing the cellular alterations leading to the earliest transformed cells, i.e., the initiated cell or incipient preneoplastic cell. Second, selenium could prevent the expression of initiated cells to form overt preneoplastic lesions.[56,57] Third, selenium could prevent the transformation of cells from a preneoplastic phenotype whose growth capabilities are stringently regulated to a neoplastic phenotype whose growth capabilities are less stringently regulated. Finally, selenium could inhibit the growth rate of established mammary tumors. The efficacy of selenium-mediated inhibition on these different stages of mammary tumorigenesis has been examined in a series of experiments. The current idea on the relative effectiveness of selenium-mediated inhibition is illustrated in the following scheme.

The results suggest that the early stages of mammary tumorigenesis, which are manifested as the induction and expression of preneoplastic hyperplastic lesions, are most sensitive to selenium-mediated inhibition. Thus, selenium (4 to 6 ppm) in the drinking water inhibited the occurrence of mammary preneoplasias by 75%.[37,40]

The inhibition of DMBA-induced mammary hyperplasias (nodule-like alveolar lesions-NLAL) was also demonstrated in organ culture.[58] A dose-related inhibition of NLAL was demonstrated for selenium given at doses of 10^{-8} to 10^{-4} M. High doses ($\geq 10^{-6}$) inhibited the frequency of DMBA-induced NLAL, whether given at the initiation or postinitiation stages. The efficacy of selenium was greatest when given during only the postinitiation stages compared to selenium given during only the initiation stage. At low doses of selenium (10^{-8}, 10^{-7} M), there was a modest increase (14 to 16%) in the percent of glands with NLAL when selenium was given at either of the two stages. These results demonstrated the direct effect of selenium at the level of the target organ and imply that systemic alterations need not be implicated in the mechanism of selenium-mediated inhibition of tumorigenesis.

The data also suggest that events involved in tumor formation from preneoplasias can also be modulated by selenium, although the degree of inhibition may not be as great as occurs in the earlier stages. Thus, of the 14 nodule outgrowth lines examined in syngenic mice treated with selenium (4 to 8 ppm) in the drinking water, only 5 of the lines (36%) exhibited a significant inhibition in tumor incidence or tumor latency.[40] Nodule outgrowth line D2 was refractory to the effects of selenium even at 8 ppm in the water.[124] In contrast, the growth of established mammary tumor populations is essentially refractory to selenium supplementation. Thus, the net growth of first and second generation transplantable mammary tumors was inhibited by selenium in less than 10% of the samples. This result occurred even under conditions where the mice were exposed to selenium (7 ppm) starting 6 weeks before the tumors were transplanted or when selenium was injected 2 μg/g body weight for 3× weekly with the tumor growing subcutaneously.[36,40]

The lack of a significant inhibitory effect of selenium on the growth rate of primary tumors was similar to the results obtained by Schrauzer and co-workers in C3H/St mice[32,33] and Ip in the DMBA-treated rat.[27] Of interest was the report of Ip et al.[59] that 2 ppm Se inhibited the growth of the transplantable rat mammary tumor MT-W9B. However, the data clearly showed an inhibition only of the early events which occurred during the establishment of a tumor transplant. The tumors in the selenium-treated rats exhibited a lag of approximately 5 days compared to tumors growing in the untreated rats. Once the tumors start growing, the growth rate of the tumors between the two groups of rats was the same. Thus, it is conceivable that selenium was interfering with events unrelated to tumor growth per se, i.e., a neo-vascularization response. Milner and co-workers[60-63] have reported that selenium significantly inhibited the in vivo growth of Ehrlich ascites tumor cells, L1210 leukemia cells, and a canine mammary tumor line. The efficacy of selenium-mediated inhibition was dependent upon the chemical form of selenium (selenite > selenate; selenite > selenocystine > selenomethionine), on the dose of selenite, and on the route of administration (i.p. > drinking water). All three tumor lines represent highly selected cell populations which have been passaged in vivo as ascites tumors or in vitro monolayer culture for many years. Thus, their response to an agent may not accurately reflect the response of the heterogeneous cell populations typical of primary tumors *in situ*[64] or tumors in early transplant generations. Such results should be interpreted with caution and be used to understand some of the mechanisms of selenium action. From published work, it is likely these model tumor systems may be represented by only a minority of primary mammary tumors.

III. INHIBITION OF CELL GROWTH IN VITRO

The use of in vitro cell culture techniques allows one to examine the direct effects of an agent on cell proliferation and differentiation. McKeehan et al.[65] were the first to demonstrate that selenium was essential for the clonal growth of cells in monolayer culture, the WI-38 cell line of normal human fibroblasts. Since then, selenium has been shown to be an essential and/or an important factor for the growth of human erythroid,[66] HeLa,[67] colon carcinoma,[68] oat cell carcinoma,[69] hepatoma,[70] and normal and neoplastic bladder cells,[71]; rat neuroblastoma and pituitary GH_3 cells,[72] and mouse melanoma,[72] 3T3,[73] and mammary cells.[74]

The effects of selenium on the growth kinetics of normal, preneoplastic, and neoplastic mammary cells grown in vitro as primary cultures and as established cell lines was examined by Medina and co-workers.[74] Selenium (10^{-5} M) present as Na_2SeO_3 in serum-free Dulbecco's MEM, inhibited the growth of all three mammary cell populations in primary culture as well as seven established mammary cell lines. At 5×10^{-6} M, selenium inhibited normal mammary cells, one of two preneoplastic cell lines, and D2 tumors in primary culture, but only one of three established mammary cell lines (YN-4). Of more interest was the observation that at 5×10^{-8} M, selenium stimulated the growth of primary cell cultures of normal mammary cells, one of two preneoplastic cell lines, and one of three established mammary cell lines, but not the growth of tumors in primary cell culture. This biphasic effect of selenium on cell growth has been also seen for WI-38 human fibroblasts,[65] Chang liver cells, and 3T3 cells.[73] A biphasic effect was observed also by Chatterjee and Banerjee[58] on the effects of selenium on DMBA-induced noduligenesis in mammary gland organ culture.

The response of D2 mammary tumors grown in primary cell cultures to 5×10^{-6} M Se was puzzling since this population appeared responsive to selective inhibition in vitro, but not in vivo. Primary D2 tumors, the same cell populations that were grown in vitro, were unresponsive to selenium-mediated inhibition when assayed by s.c. transplantation in syngeneic mice and treated with selenium[40] under the same conditions that Milner and co-workers[60-62] had successfully shown inhibition of tumor growth. These results reinforce the caution raised earlier on extending results generated from highly selected cell populations like L1210 leukemia to conditions found in *in situ* neoplasms. D2 primary tumors, grown in primary culture, behave as partly synchronized cell populations; thus they proceed through two waves of cell division characterized by a high proliferative compartment. However, primary mammary tumors even when serially transplanted for two to three transplant generations, are a heterogeneous population with a relatively small proliferative compartment.

Thus, L1210 cells are more analogous to D2 and YN-4 mammary cells in vitro than to D2 tumor cells in vivo. It is important to recognize that L1210 leukemia cells and CMT-14B canine mammary cells in vivo and YN-4 mammary cells in vitro are useful model systems to examine the mechanisms of action of selenium, but the results cannot be used to support the hypothesis that selenium effectively inhibits the growth of primary tumors *in situ.*

The growth inhibitory effects of selenium are reversible. This reversibility was seen for mammary cells in organ and cell culture,[58,75] for HeLa S_3 cells,[76] and liver cells in cell culture.[77] The reversibility was dose dependent since the growth of mammary cells exposed to $5 \times 10^{-6} M$ Se, but not $5 \times 10^{-5} M$ Se was reversible. The lag time preceding recovery may be associated with the time necessary for selenium to be cleared from the cell since experiments for mammary cells indicated the half-life for intercellular selenium was 2.6 days.[78] The observations in vitro coincided with observations in vivo which indicated that inhibition of mammary tumorigenesis was dependent upon the continued presence of selenium.[28,29,35]

In summary, the in vitro model systems provide useful systems to examine the direct effect of selenium on mammary epithelial cells and determine the possible mechanisms of action.

IV. MECHANISMS OF ACTION

A. Antioxidant

The only function characterized for selenium in mammalian cells is its antioxidant function.[9,79,80] The enzyme, GSH-Px, is a seleno-enzyme, which functions as a lipid antioxidant. GSH-Px, along with vitamin E, functions to eliminate organic peroxides from the cell. Whereas vitamin E is located primarily in cell membranes, GSH-Px is a soluble enzyme which is present in the cytosol and mitochondrial matrix. With this information, it was natural to examine the interrelationship between selenium and GSH-Px in mammary tumorigenesis.

Lane et al.[11,81] examined the levels of selenium and GSH-Px in the normal and neoplastic developmental states of the mammary gland. The enzyme in the mammary gland was inducible by dietary selenium concentrations from 0.02 to 0.10 ppm but was not further increased when dietary selenium was above 0.10 ppm. The biochemical characteristics of the mammary GSH-Px enzyme were similar to that reported for other tissues. In mice fed a chow diet without supplemental selenium (0.15 ppm Se), the levels of mammary GSH-Px and selenium increased in pregnant and lactating mice compared to virgin mice. This increase paralleled the increase in the epithelial component of the mammary gland. Interestingly, preneoplastic and neoplastic mammary tissues contained GSH-Px levels similar to pregnant mammary gland, whereas selenium levels in the tumor tissues were 50% greater than pregnant tissues. As discussed in an earlier section (II.B), these tumor tissues were unresponsive to selenium-mediated inhibition of growth, even though the selenium concentration reached 0.5 ppm in these tissues. Whereas the selenium concentration in the mammary gland was responsive to the levels of dietary selenium at each stage of mammary gland development, the levels of GSH-Px were responsive only in those mammary tissues exhibiting a rapid growth rate (i.e., mammary GSH-Px in 10- but not 26-week-old virgin mice was responsive to dietary selenium levels). Ip and co-workers reached a similar conclusion in their studies on rat mammary tumorigenesis.[31,51] Thus, in the DMBA-treated rat, selenium supplement failed to increase the basal glutathione peroxidase activity in the mammary gland. A somewhat similar finding was reported by Jacobs et al.[20] on the levels of GSH-Px and selenium in DMH-treated rats. Whereas the levels of liver selenium increased with increasing selenium intake, the levels of GSH-Px actually decreased. Unfortunately, DMH induces colonic tumors and the enzyme and selenium levels were not measured in colon tissues. In summary, the above results support the conclusion that the inhibitory effects of high levels of selenium are not related to its biochemical regulation of the selenium-dependent glutathione peroxidase. Apparently, the enzyme is operating at maximal capacity at physiological levels of selenium. Furthermore, there seems to be no direct correlation between tissue selenium levels and inhibition of tumor cell growth.[11]

A second approach to examine a related mechanism was to measure the effect of selenium on lipid peroxidation in the mammary gland. Ip and Sinha[82] found no correlation between the anticarcinogenic effect of selenium supplementation and its ability to suppress lipid peroxidation in mammary tissues, regardless whether the rats were fed a diet high in saturated or unsaturated fat. Although the assay method used in this study, the thiobarbituric acid (TBA) method, was an indirect measurement of lipid peroxidation, it was probably sufficiently sensitive to pick up major alterations. Additionally, Horvath and Ip[31] demonstrated that vitamin E suppressed lipid peroxidation in the mammary gland under conditions where it did not inhibit mammary tumorigenesis. However, before this mechanism is completely ruled out, the experiments of Wong et al.[83] should be considered. They provided evidence that N-OH-AAF was activated by a lipoxygenase peroxidase route and that this enzyme was inhibited by selenium. Unfortunately, the details of the experiment were not provided. In summary, the majority of the experiments so far indicate that the anticarcinogenic effect of selenium is not correlated with GSH-Px levels or its presumed antioxidant function; however, it is important to examine other peroxidative pathways and also use more direct methods to measure lipid peroxidation before this potential pathway for selenium action can be eliminated from consideration.

B. Cytological Events

Although GSH-Px is the only characterized seleno-enzyme, there are other presumptive selenoproteins in the mammalian cells.[80] The functions of these proteins are unknown, although they have been associated with selenium deficiency diseases, such as white muscle disease[84] and sperm immotility,[85,86] and identified as transport proteins[87,88] and as Se-labeled proteins in testis,[89] kidney,[90,91] and mammary gland.[92] Until the functional properties and cellular compartmentalization of these proteins are better understood, they provide only a lead to examine other modes of selenium action. However, some limited studies have examined the effect of selenium on cytological alterations in mammary cells under conditions where selenium inhibits cell growth.[93,94] Briefly, selenium did not alter any component of the fibrillar cytoskeletal system, but it did lead to inclusions in the mitochondrial matrix of responsive mammary cells. It has been known for 30 years that high doses of selenium (10^{-5} M) inhibit the activity of mitochondrial succinate dehydrogenase.[95,96] Furthermore, a considerable amount of selenium (20 to 25%) is localized within mitochondria of liver and mammary epithelial cells.[78,97] The nature of the mitochondrial inclusions are unknown. These inclusions could not be correlated with changes in the activity of the mitochondrial enzymes, cytochrome C oxidase and succinate dehydrogenase, thus implying there was no demonstrable impairment of mitochondria under these conditions.[94] At this time, information on the cytological effects of selenium under conditions where it inhibits tumorigenesis is scarce and more investigation into membrane and nuclear changes in defined model systems is warranted.

C. Selenium-Carcinogen Interactions

Numerous chemical carcinogens are potent mutagens in the Ames' Salmonella mutagenesis assay.[98] Since selenium inhibits both direct- and indirect-acting carcinogens, several investigators have examined the effect of selenium on the inhibition of carcinogen-induced mutagenicity. These and other studies have generated an area of confusion concerning the potential antimutagenic and mutagenic properties of selenium.[99-108] The evidence for the inhibitory effects of selenium (as selenite) on the mutagenicity of carcinogens is provided by only a few reports. Jacobs et al.[100] demonstrated that selenium decreased to 65% of controls the mutagenicity of AAF, N-OH-AAF, and N-OH-AF in Salmonella test strains. The effective molar ratio of Se/AAF was 10:1 but the required dose of selenium was high (10 mM). Furthermore, it was unclear if the cytotoxic properties of selenium were assessed and controlled in these experiments. Shamberger et al.[102] demonstrated that selenium at low doses (0.0067 → 0.67 μmol) inhibited frame shift mutagenesis induced by the carcinogens malonaldehyde and β-propiolactone in 5 Salmonella tester strains. Rosin and Stitch[107] also demonstrated that selenium inhibited MNNG and acetoxy-AAF induced mutagenesis in the Ames assay. Furthermore, Rosin[105] showed that selenium inhibited spontaneous mutagenesis in yeast cultures under conditions where the growth rate of the yeast was not altered. Many metal salts which are mutagenic will affect the fidelity of DNA synthesis. However, whereas chromium, a known carcinogen altered the fidelity of DNA synthesis, selenium had no effect under the same assay conditions.[108]

In contrast to these observations, there are several observations in the literature which show that selenium increases mutagenesis in human leukocytes,[99] induces DNA fragmentation,[101] DNA repair synthesis,[101,103] chromosome aberrations in fibroblasts,[99] and sister chromatid exchange in human leukocytes[104] and Chinese hamster V79 cells. The confusing results are understandable if the doses of selenium used in the various experiments are considered. In general, the antimutagenic properties of selenium are evident under conditions where selenium does not exhibit a toxic effect ($\leq 5 \times 10^{-6}$ M). The mutagenic and deleterious cytogenic effects are seen where selenium has a toxic effect, (i.e., $\geq 10^{-5}$ M).[1] Thus, in any consideration about the mechanism of action of selenium, it is vital that the concentrations (as well as the form of selenium) be considered in evaluating the particular parameters under examination.

Since selenium inhibits the initiation phase of carcinogenesis, it is important to determine if selenium alters carcinogen metabolism or carcinogen-DNA interactions. Although only a few experiments have been performed in these areas, the results suggest the areas where positive interactions may occur. Selenium inhibited AHH activity by 50% in lymphocytes over a range of benzpyrene concentrations.[109] However, the dose of selenium necessary to inhibit AHH activity was 100 μM; thus, the specificity of the effect is debatable. In another experiment, selenium (4 ppm) in the drinking water resulted in an increase in ring hydroxylation and a decrease in N-hydroxylation of AAF.[16] The increase in 3-OH-AAF and a decrease in N-OH-AAF was also seen by adding selenium to a microsomal assay system. In a similar experiment, selenium enhanced liver glucuronyl transferance activity and decreased sulfotransferase activity.[110] Thus, these experiments indicated that selenium can shift the balance of carcinogen metabolism towards detoxification pathways, thereby reducing the risk of carcinogen-induced cellular alterations which lead to neoplastic transformation. Selenium also decreased hepatic metabolism of DMH.[111]

The influence of selenium on carcinogen binding to DNA was reported by several investigators. Low levels of dietary selenium (0.5 ppm) did not alter AAF covalently bound to liver DNA;[112] however, higher levels (4 ppm) given in the drinking water, decreased the binding of labeled AAF to liver DNA.[110] In a third study, selenium did not influence the binding to DMH to colonic DNA.[113]

The influence of selenium on carcinogen-DNA adduct formation has been reported by two investigators. Bull et al.[113] found that selenium had no effect on O^6-methylguanine formation in DMH-treated colon, whereas Harbach and Swenberg[111] found that selenium increased N-7 and O^6-methylguanine formation in the colon of DMH-treated rats. However, selenium inhibited the formation of these two adducts in the livers of such rats. Since the colon is the target organ for DMH-carcinogenesis, these experiments provided no support for the idea that selenium intervened at the level of DNA-adduct formation. In contrast to these findings, selenium increased the repair rate of BOP-induced DNA damage in the colon and AAF-induced DNA damage in the liver.[112] Thus, a consideration of the published results suggests that selenium might ameliorate the effects of carcinogens during the initiation process by altering carcinogen metabolism and/or enhancing repair of carcinogen-induced DNA breaks. There is little evidence that selenium alters carcinogen-DNA binding or DNA alkylation events.

D. Inhibition of Cell Proliferation

Numerous experiments have implicated selenium as an important nutrient for the growth of cells in vitro. Conversely, Milner and co-workers demonstrated that selenium supplementation inhibited the growth of several neoplastic cell lines.[60-62] On the basis of his results, Milner suggested that selenium exerted a marked and specific effect on cell proliferation.[61] Harbach and Swenberg[111] suggested that selenium inhibition of DMH-induced colonic tumorigenesis was attributable to a decrease in DNA synthesis since ^{3}H-thymidine incorporation was reduced 65%. Gruenwedel and Cruickshank[76] demonstrated that 5 μM Se inhibited the incorporation of labeled precursors into DNA and RNA of HeLa cells; protein synthesis was slightly less sensitive by a factor of 2. Interestingly, 1 mM Se was necessary before the plasma membrane became permeable to trypan blue.

The effects of selenium on DNA synthesis in a mammary cell line grown in vitro was examined by Medina et al.[74,75] As discussed above (Section III), low doses of selenium stimulated cell growth, whereas high doses (5 $\times$ 10^{-6} M) inhibited cell growth reversibly. The decreased cell growth was reflected by a decreased cell number, decreased uptake of ^{3}H-thymidine into DNA, decreased DNA labeling index, and a decreased rate of DNA synthesis. The increased cell growth was reflected by an increase in all these parameters of cell growth kinetics. Flow cytofluorometry analysis of selenium-treated mammary cells demonstrated that the cells appeared to be delayed transiently in G_2 and delayed for a prolonged time in the S-phase of the cell cycle.[94] In contrast, mammary cells exposed to 5 $\times$ 10^{-8} M Se appeared to continue to transverse the cell cycle at a time when the control cells were entering plateau conditions. The sum *toto* of the above experiments indicated that one of the mechanisms of selenium modulation of carcinogenesis could be through inhibition of DNA synthesis. The molecular basis for this effect remains to be determined.

The effects of selenium and its metabolic products on protein synthesis have been examined in several experiments. Selenium, at nanomolar levels, was a potent inhibitor of polyribosomal amino acid incorporation.[114,115] Inhibition was dependent upon thiols, i.e., glutathione (GSH), in the reaction mixture. Vernie et al.[115,116] demonstrated that selenodiglutathione (GSSeSG), a reaction product between selenium and GSH, was the active agent and inhibited protein synthesis by inhibiting elongation factor 2. Protein synthesis could be inhibited also in 3T3 cells by GSSeSG under conditions where DNA and RNA synthesis was not inhibited.[117] Protein synthesis was not inhibited in *E. coli* nor was elongation factor G blocked by selenium. Interestingly, the inhibitory effect could not be reversed by glutathione.[116,117] Vernie et al.[118] also demonstrated that GSSeSG injected i.p. inhibited tumor growth and lifespan of mice previously inoculated with mouse malignant lymphocytes. Although GSSeSG killed up to 99% of the cells recovered from the peritoneal fluid, the survival time was increased from only 11 days in control mice to 14 days in treated mice and death did not seem to correlate with tumor cell number in treated mice. The concentration of GSSeSG (0.83 μmol) was not toxic to nontumor bearing mice; therefore, the relationship between the strong inhibition of protein synthesis and the weak inhibition of tumor cell growth was difficult to understand.

Selenium, administered s.c. or i.p., was shown to alter selectively cAMP metabolism in a hepatoma cell line.[119] Selenium increased cAMP levels in tumor cells, but not in the host normal liver cells, by decreasing phosphodiesterase (PDE) activity. The basis for this selectivity was related to differences in PDE isoenzyme patterns. The effects on selenium on hepatoma cell growth was not noted in this study.

The above results demonstrate that selenium can selectively effect macromolecular synthesis at doses which inhibit cell growth. Further studies along these lines might clarify the molecular basis for the ability of selenium to inhibit cell growth.

E. Other Modes

An intriguing mechanism for the effect of selenium was proposed by Spallholz et al.,[120,121] who suggested that selenium acts as an immunoadjuvant. In support of this hypothesis were the observations that selenium enhanced the primary immune response in mice by increasing the number of plaque-forming cells (PFC) and agglutinating antibody, and promoting immunoglobulins M and G synthesis. These effects were seen at relatively low levels of dietary selenium (0.75 to 2.0 ppm). A similar finding was reported by Shakelford and Martin[122] when selenium was given in the drinking water. Martin and Spallholz[123] also demonstrated that dietary selenium enhanced cell-mediated immune response. The immunostimulatory effects of selenium have also been demonstrated in several other species.[1] So far, there do not appear to be any experiments where the immunoregulatory effects of selenium on tumors or in a carcinogenesis study have been examined.

Since mammary tumorigenesis is markedly influenced by the hormonal status of the host, it is worthwhile to know if selenium alters the hormonal milieu of the host. Ip[27] reported there was no difference in the estrus cycle of the rat, regardless of selenium intake (0.1 to 2.5 ppm). Furthermore, prolactin, estrone, and estradiol serum levels determined at proestrus were the same for rats fed 0.1 or 2.5 ppm dietary Se. Ip[28] also showed that selenium delayed the regrowth of tumors that had regressed after ovariectomy. The above two approaches which examined the effects of selenium on host systemic events indicate that there is little evidence available to suggest an indirect effect for selenium-mediated inhibition of tumorigenesis.

V. CONCLUSION

The impressive amount of experiments over the past 10 years has clearly documented the efficacy of selenium as an inhibitor of the neoplastic transformation in a variety of epithelial organs. Although the principal physiological role of selenium in vertebrate cell physiology is explainable by the enzyme GSH-Px, secondary functions in normal cell physiology as well as the function of selenium as an inhibitor of the neoplastic transformation are not understood and experiments to examine these functions have only just begun. It is important to pursue rigorously multiple approaches to the understanding of the mechanism of action of selenium. First, further animal studies are warranted to assess the interaction of selenium with other trace elements, vitamins, and inhibitory factors from natural products in order to develop a rationale approach for the use of dietary components to decrease the risk of cancer. The ultimate goal of a "nutritional prevention" of cancer is a valid one; however, the fulfillment of this goal is still distant. Second, the availability of well-developed model systems in the colon and mammary gland should facilitate experiments on the function of selenium at the cell and molecular level. Dietary selenium supplementation inhibits both the initiation and postinitiation phases of the neoplastic process and different mechanisms may be operative at each stage. Third, the existence of other physiological functions for selenium has been suggested repeatedly in the scientific literature; thus, experiments addressed to this aim are highly warranted.

Finally, it is important to recognize that the role of selenium in human physiology is incompletely understood. Selenium can be a very toxic substance at high concentrations, thus, it is important to obtain more information on human tolerance levels, bioavailability of selenium in different foods, and appropriate epidemiological information on the possible relationships between human selenium exposure and cancer risk.

REFERENCES

1. **Shamberger, R. J.,** *Biochemistry of Selenium,* Plenum Press, New York, 1983.
2. **Wilbur, C. G.,** Toxicology of selenium: a review, *Clin. Toxicol.,* 17, 171, 1980.
3. **Franke, K. W.,** A new toxicant occurring naturally in certain samples of plant foodstuffs, *J. Nutr.,* 8, 597, 1934.
4. **Martin, J. L.,** Selenium assimilation in animals, in *Organic Selenium Compounds: Their Chemistry and Biology,* Klayman, D. L. and Gunther, W. H., Eds., John Wiley & Sons, New York, 1973, 663.
5. **Schwartz, K. and Foltz, C. M.,** Selenium as an integral part of Factor 3 against dietary necrotic liver degeneration, *J. Am. Chem. Soc.,* 79, 3292, 1957.
6. **Muth, O. H., Oldfield, J. E., Remmert, L. F., and Shubert, J. R.,** Effects of selenium and vitamin E on white muscle disease, *Science,* 128, 1090, 1958.
7. **Thompson, J. N. and Scott, M. L.,** Impaired lipid and vitamin E adsorption related to atrophy of the pancreas in selenium-deficient chicks, *J. Nutr.,* 100, 797, 1970.
8. **Maag, D. D. and Glenn, M. W.,** Toxicity of selenium: farm animals, in *Selenium in Biomedicine,* Muth, O. H., Oldfield, J. E., and Weswig, P. H., Eds., AVI, Westport, Conn., 1967, 127.
9. **Rotruck, J. T., Pope, A. L., Ganther, H. E., Hafeman, D. G., and Hoekstra, W. G.,** Selenium: biochemical role as a component of glutathione peroxidase, *Science,* 179, 588, 1973.
10. **Oh, S. H., Ganther, H. E., and Hoekstra, W. G.,** Selenium as a component of glutathione peroxidase isolated from ovine erythrocytes, *Biochemistry,* 13, 1823, 1974.
11. **Lane, H. W. and Medina, D.,** Selenium concentration and glutathione peroxidase activity in normal and neoplastic development of the mouse mammary gland, *Cancer Res.,* 43, 1558, 1983.
12. **Nelson, A. A., Fitzhugh, O. G., and Calvery, H. O.,** Liver tumors following cirrhosis caused by selenium in rats, *Cancer Res.,* 3, 230, 1943.
13. **Clayton, C. C. and Baumann, C. A.,** Diet and azo dye tumors: effect of diet during a period when the dye is not fed, *Cancer Res.,* 9, 575, 1949.
14. **Harr, J. R., Exon, J. H., Weswig, P. H., and Whanger, P. D.,** Relationship of dietary selenium concentration, chemical cancer induction, and tissue concentration of selenium in rats, *Clin. Toxicol.,* 5, 287, 1972.
15. **Griffin, A. C. and Jacobs, M. M.,** Effects of selenium on azo dye hepatocarcinogenesis, *Cancer Lett.,* 3, 177, 1977.
16. **Marshall, M. W., Arnott, M. S., Jacobs, M. M., and Griffin, A. C.,** Selenium effects on the carcinogenicity and metabolism of 2-acetylaminofluorene, *Cancer Lett.,* 7, 331, 1979.
17. **Griffin, A. C.,** Role of selenium in the chemoprevention of cancer, *Adv. Cancer Res.,* 29, 419, 1979.
18. **Jacobs, M. M., Jansson, B., and Griffin, A. C.,** Inhibitory effects of selenium on 1,2-dimethylhydrazine and methylazoxymethanol acetate induction of colon tumors, *Cancer Lett.,* 2, 133, 1977.
19. **Soullier, B. K., Wilson, P. S., and Nigro, N. D.,** Effect of selenium on azomethane-induced intestinal cancer in rats fed high fat diet, *Cancer Lett.,* 12, 343, 1981.
20. **Jacobs, M. M., Frost, C. F., and Beams, F. A.,** Biochemical and clinical effects of selenium on dimethylhydrazine-induced colon cancer in rats, *Cancer Res.,* 41, 4458, 1981.
21. **Birt, D. F., Lawson, T. A., Julius, A. D., Runice, C. E., and Salmasi, S.,** Inhibition by dietary selenium of colon cancer induced in the rat by bis(2-oxoxpropyl)nitrosamine, *Cancer Res.,* 42, 4455, 1982.
22. **Ankerst, J. and Sjogren, H. O.,** Effect of selenium on the induction of breast fibroadenomas by adenovirus type 9 and 1,2-dimethylhydrazine-induced bowel carcinogenesis in rats, *Int. J. Cancer,* 29, 707, 1982.
23. **Jacobs, M. M.,** Selenium inhibition of 1,2-dimethylhydrazine-induced colon carcinogenesis, *Cancer Res.,* 43, 1646, 1983.
24. **Thompson, H. J. and Becci, P. J.,** Selenium inhibition of N-methyl-N-nitrosourea-induced mammary carcinogenesis in the rat, *J. Natl. Cancer Inst.,* 65, 1299, 1980.
25. **Welsch, C. W., Goodrich-Smith, M., Brown, C. K., Greene, H. D., and Hamel, E. J.,** Selenium and the genesis of murine mammary tumors, *Carcinogenesis,* 2, 519, 1981.
26. **Thomson, H. J., Meeker, L. D., and Becci, P. J.,** Effect of combined selenium and retinyl acetate treatment on mammary carcinogenesis, *Cancer Res.,* 41, 1413, 1981.

27. **Ip, C.**, Factors influencing the anticarcinogenic efficacy of selenium in dimethylebenz[a]anthracene-induced mammary tumorigenesis in rats, *Cancer Res.*, 41, 2683, 1981.

28. **Ip, C.**, Prophylaxis of mammary neoplasia by selenium supplementation to the initiation and promotion phases of chemical carcinogenesis, *Cancer Res.*, 41, 4386, 1981.

29. **Ip, C. and Ip, M. M.**, Chemoprevention of mammary tumorigenesis by a combined regimen of selenium and vitamin A, *Carcinogenesis*, 2, 915, 1981.

30. **Thompson, H. J., Meeker, L. D., Becci, P. J., and Kokoska, S.**, Effect of short-term feeding of sodium selenite on 7,12-dimethylbenz[a]anthracene-induced mammary carcinogenesis in the rat, *Cancer Res.*, 42, 4954, 1982.

31. **Horvath, P. M. and Ip, C.**, Synergistic effect of vitamin E and selenium in the chemoprevention of mammary carcinogenesis in rats, *Cancer Res.*, 43, 5335, 1983.

32. **Schrauzer, G. N. and Ishmael, D.**, Effects of selenium and arsenic on the genesis of spontaneous mammary tumors in inbred C3H mice, *Ann. Clin. Lab. Sci.*, 4, 441, 1974.

33. **Schrauzer, G. N., White, D. A., and Schneider, C. J.**, Inhibition of the genesis of spontaneous mammary tumors in C3H mice: effects of selenium and of selenium-antagonistic elements and their possible role in human breast cancer, *Bioinorg. Chem.*, 6, 265, 1976.

34. **Schrauzer, G. N., White, D. A., and Schneider, C. J.**, Selenium and cancer: effects of selenium and of the diet on the genesis of spontaneous mammary tumors in virgin inbred female C3H/St mice, *Bioinorg. Chem.*, 8, 387, 1978.

35. **Schrauzer, G. N., McGinness, J. E., and Kuehn, K.**, Effects of temporary selenium supplementation on the genesis of spontaneous mammary tumors in inbred female C3H/St mice, *Carcinogenesis*, 1, 199, 1980.

36. **Medina, D. and Shepherd, F.**, Selenium-mediated inhibition of mouse mammary tumorigenesis, *Cancer Lett.*, 8, 241, 1980.

37. **Medina, D. and Shepherd, F.**, Selenium-mediated inhibition of 7,12-dimethylbenz[a]anthracene-induced mouse mammary tumorigenesis, *Carcinogenesis*, 2, 451, 1981.

38. **Medina, D., Lane, H. W., and Shepherd, F.**, Effects of selenium on mouse mammary tumorigenesis and glutathione peroxidase activity, *Anticancer Res.*, 1, 377, 1981.

39. **Medina, D., Lane, H. W., and Shepherd, F.**, Effect of dietary selenium levels on 7,12-dimethylben-zanthracene-induced mouse mammary tumorigenesis, 4, 1159, 1983.

40. **Medina, D. and Lane, H. W.**, Stage specificity of selenium-mediated inhibition of mouse mammary tumorigenesis, *Biol. Trace Element Res.*, 5, 297, 1983.

41. **Shamberger, R. J.**, Relationship of selenium to cancer. 1. Inhibitory effect of selenium on carcinogenesis, *J. Natl. Cancer Inst.*, 44, 931, 1970.

42. **Thompson, H. J. and Becci, P. J.**, Effect of graded dietary levels of selenium on tracheal carcinomas induced by 1-methyl-1-nitrosourea, *Cancer Lett.*, 7, 215, 1979.

43. **Shamberger, R. J., Tytko, S. A., and Willis, C. E.**, Antioxidants and cancer. VI. Selenium and age-adjusted human cancer mortality, *Arch. Environ. Health*, 31, 231, 1976.

44. **Shamberger, R. J., Tytko, S. A., and Willis, C. E.**, Antioxidants in cereals and in food preservatives and declining gastric cancer mortality, *Cleveland Clin. Q.*, 39, 199, 1972.

45. **Schrauzer, G. N., White, D. A., and Schneider, C. J.**, Cancer mortality correlation studies. III. Statistical associations with dietary selenium intakes, *Bioinorg. Chem.*, 7, 23, 1977.

46. **Jansson, B. and Jacobs, M. M.**, Selenium — a possible inhibitor of colon and rectum cancer, in *Proc. Symp. Selenium — Tellurium in the Environment*, Industrial Health Foundation, Pittsburgh, 1976, 326.

47. **Shamberger, R. J., Rukovera, E., Longfield, A. K., Tytko, S. A., Deodhar, S., and Willis, C. E.**, Antioxidants and cancer. I. Selenium in the blood of normals and cancer patients, *J. Natl. Cancer Inst.*, 50, 863, 1973.

48. **McConnell, K. P., Broghamer, W. L., Blotcky, A. J., and Hurt, O. J.**, Selenium levels in human blood and tissues in health and disease, *J. Nutr.*, 105, 1026, 1975.

49. **McConnell, K. P., Jagar, R. M., Higgins, P. J., and Blotcky, A. J.**, Serum selenium levels in patients with and without breast cancer, in *Proc. 18th Meet. Am. Coll. Nutr.*, Van Eys, J., Seelig, M. S., and Nichols, B. L., Eds., S. P. Medical and Scientific Books, New York, 1977, 195.

50. **Goodwin, W. J., Lane, H. W., Bradford, K., Marshall, M., Griffin, A. C., Geofert, H., and Jesses, R. H.**, Selenium and glutathione peroxidase levels in epidermoid carcinoma of the oral cavity and oropharynx, *Cancer*, 51, 110, 1983.

51. **Ip, C. and Sinha, D. K.**, Enhancement of mammary tumorigenesis by dietary selenium deficiency in rats with a high polyunsaturaed fat intake, *Cancer Res.*, 41, 31, 1981.

52. **DeOme, K. B., Faulkin, L. J., Jr., Bern, H. A., and Blair, P. B.**, Development of mammary tumors from hyperplastic alveolar nodules transplanted into gland-free mammary fat pads of female C3H mice, *Cancer Res.*, 19, 515, 1959.

53. **Medina, D.**, Preneoplastic lesions in mouse mammary tumorigenesis, *Methods Cancer Res.*, 7, 3, 1973.

54. **Cardiff, R. D., Wellings, S. R., and Faulkin, L. J., Jr.,** Biology of breast neoplasia, *Cancer,* 39, 2734, 1977.
55. **Medina, D. and Asch, B. B.,** Cell markers for mouse mammary preneoplasias, in *Cell Biology of Breast Cancer,* McGrath, C. M., Brennan, M. J., and Rich, M. A., Eds., Academic Press, New York, 1980, 363.
56. **DeOme, K. B., Miyamoto, M. J., Osborn, R. C., Guzman, R. C., and Lum, K.,** Detection of inapparent nodule-transformed cells in the mammary gland tissues of virgin female BALB/cfC3H mice, *Cancer Res.,* 38, 2102, 1978.
57. **Guzman, R. C., Osborn, R. C., and DeOme, K. B.,** Recovery of transformed nodule and ductal mammary cells from carcinogen-treated C57BL mice, *Cancer Res.,* 47, 1808, 1981.
58. **Chatterjee, M. and Banerjee, M. R.,** Selenium-mediated dose inhibition of 7,12-dimethylbenzanthracene-induced transformation of mammary cells in organ culture, *Cancer Lett.,* 17, 187, 1982.
59. **Ip, C., Ip, M. M., and Kim, U.,** Dietary selenium intake and growth of the MT-W9B transplantable rat mammary tumor, *Cancer Lett.,* 14, 101, 1981.
60. **Greeder, G. A. and Milner, J. A.,** Factors influencing the inhibitory effect of selenium on mice inoculated with Ehrlich ascites tumor cells, *Science,* 209, 825, 1980.
61. **Milner, J. A. and Hsu, C. Y.,** Inhibitory effects of selenium on the growth of L1210 leukemic cells, *Cancer Res.,* 41, 1652, 1981.
62. **Watrach, A. M., Milner, J. A., and Watrach, M. A.,** Effect of selenium on growth rate of canine mammary carcinoma cells in athymic nude mice, *Cancer Lett.,* 15, 137, 1982.
63. **Poirier, K. A. and Milner, J. A.,** Factors influencing the antitumorigenic properties of selenium in mice, *J. Nutr.,* 113, 2147, 1983.
64. **Heppner, G. H., Dexter, D. L., DeNucci, T., Miller, F. R., and Calabresi, P.,** Heterogeneity in drug sensitivity among tumor cell subpopulations of a single mammary tumor, *Cancer Res.,* 38, 3758, 1978.
65. **McKeehan, W. L., Hamilton, W. G., and Ham, R. G.,** Selenium is an essential trace element for growth of WI-38 diploid human fibroblasts, *Proc. Natl. Acad. Sci. U.S.A.,* 73, 2023, 1976.
66. **Guilbert, L. J. and Iscove, N. N.,** Partial replacement of serum by selenite, transferrin, albumin, and lecithin in haemapoietic cell cultures, *Nature (London),* 263, 594, 1976.
67. **Hutchings, S. E. and Sato, G. H.,** Growth and maintenance of HeLa cells in serum-free medium supplemented with hormones, *Proc. Natl. Acad. Sci. U.S.A.,* 75, 901, 1978.
68. **Murakami, H. and Masui, H.,** Hormonal control of human colon carcinoma cell growth in serum-free medium, *Proc. Natl. Acad. Sci. U.S.A.,* 77, 3464, 1980.
69. **Simms, E., Gazdar, A. F., Abrams, P. G., and Minna, J. D.,** Growth of human small cell (oat cell) carcinoma of the lung in serum-free growth factor supplemented medium, *Cancer Res.,* 40, 4356, 1980.
70. **Nakabayashi, H., Taketa, K., Miyano, K., Yamane, T., and Sato, J.,** Growth of human hepatoma cell lines with differentiated functions in chemically-defined medium, *Cancer Res.,* 42, 3858, 1982.
71. **Messing, E. M., Fahey, J. J., deKernion, J. B., Bhuta, S. M., and Bubbers, J. E.,** Serum-free medium for the *in vitro* growth of normal and malignant bladder epithelial cells, *Cancer Res.,* 42, 2392, 1982.
72. **Bottenstein, J. E., Hayashi, I., Hutchings, S., Masui, H., Mather, J., McClure, D. B., Ohasa, S., Rizzina, A., Sato, G. H., Serrero, G., Wolfe, R., and Wu, B.,** The growth of cells in serum-free hormone-supplemented media, *Methods Enzymol.,* 58, 94, 1979.
73. **Potter, S. D. and Matrone, G.,** A tissue culture model for mercury-selenium interactions, *Toxicol. Appl. Pharmacol.,* 40, 201, 1977.
74. **Medina, D. and Oborn, C. J.,** Differential effects of selenium on the growth of mouse mammary cells *in vitro, Cancer Lett.,* 13, 333, 1981.
75. **Medina, D. and Oborn, C. J.,** Selenium inhibition of DNA synthesis in mammary epithelial cell line YN-4, *Cancer Res.,* 44, 4361, 1984.
76. **Gruenwedel, D. W. and Cruickshank, M. K.,** The influence of sodium selenite on the viability and intracellular synthetic activity (DNA, RNA, and protein synthesis) of the HeLa cells, *Toxicol. Appl. Pharmacol.,* 50, 1, 1979.
77. **LeBouef, R. A. and Hoekstra, W. G.,** Changes in cellular glutathione levels: possible relation to selenium-mediated anticarcinogenesis, *Fed. Proc.,* 44, 2563, 1985.
78. **Medina, D., Lane, H. W., and Oborn, C. J.,** Uptake and localization of selenium-75 in mammary epithelial cell lines *in vitro, Cancer Lett.,* 15, 301, 1982.
79. **Sunde, R. A. and Hoekstra, W. G.,** Structure, synthesis, and function of glutathione peroxidase, *Nutr. Rev.,* 38, 265, 1980.
80. **Stadtman, T. C.,** Selenium-dependent enzymes, *Ann. Rev. Biochem.,* 49, 93, 1980.
81. **Lane, H. W., Tracey, C. K., and Medina, D.,** Growth, reproduction rates and mammary gland selenium concentration and glutathione-peroxidase activity of BALB/c female mice fed two dietary levels of selenium, *J. Nutr.,* 114, 323, 1984.
82. **Ip, C. and Sinha, D.,** Anticarcinogenic effect of selenium in rats treated with dimethylbenzanthracene and fed different levels and types of fat, *Carcinogenesis,* 2, 435, 1981.

83. **Wong, P. K., Hampton, M. J., and Floyd, R. A.,** Evidence for lipoxygenase-peroxidase activation of N-hydroxy-2-acetylaminofluorene by rat mammary gland parenchymal cells, in *Prostaglandins and Cancer, 1st Int. Conf.,* Alan R. Liss, New York, 1982, 167.

84. **Pedersen, N. D., Whanger, P. D., Weswig, P. H., and Muth, O.,** Selenium binding proteins in tissues of normal and selenium responsive myopathic lambs, *Bioinorg. Chem.,* 2, 33, 1972.

85. **Calvin, H. I.,** Selective incorporation of selenium-75 into a polypeptide of the rat sperm tail, *J. Exp. Zool.,* 204, 445, 1978.

86. **Pallini, V. and Bacci, E.,** Bull sperm selenium is bound to a structural protein of mitochondria, *J. Submicr. Cytol.,* 11, 165, 1979.

87. **Burk, R. F. and Gregory, P. E.,** Some characteristics of ^{75}Se-P, a selenoprotein found in rat liver and plasma, and comparison of it with glutathione peroxidase, *Arch. Biochem. Biophys.,* 213, 73, 1982.

88. **Motsenbocker, M. A. and Tappel, A. L.,** A selenocysteine-containing selenium transport protein in rat plasma, *Biochem. Biophys. Acta,* 719, 147, 1982.

89. **McConnell, K. P., Burton, R. M., Kute, T., and Higgins, P. J.,** Selenoproteins from rat testis cytosol, *Biochem. Biophys. Acta,* 588, 113, 1979.

90. **Black, R. S., Tripp, M. J., Whanger, P. D., and Weswig, P. H.,** Selenium proteins in ovine tissues. III. Distribution of selenium and glutathione peroxidases in tissue cytosols, *Bioinorg. Chem.,* 8, 161, 1978.

91. **Motsenbocker, M. A. and Tappel, A. L.,** Selenium and selenoproteins in the rat kidney, *Biochem. Biophys. Acta,* 709, 160, 1982.

92. **Danielson, K. G., Oborn, C. J., and Medina, D.,** Analysis of selenium binding proteins in mouse mammary cells by two-dimensional gel electrophoresis, *J. Cell Biol.,* 97, 331a, 1983.

93. **Medina, D., Miller, F., Oborn, C. J., and Asch, B. B.,** Mitochondrial inclusions in selenium-treated mouse mammary epithelial cell lines, *Cancer Res.,* 43, 2100, 1983.

94. **Medina, D., Lane, H. W., and Tracey, C. M.,** Selenium and mouse mammary tumorigenesis: an investigation of possible mechanisms, *Cancer Res.,* 43, 2460, 1983.

95. **Klug, H. L., Moxon, A. L., Pedersen, D. F., and Painter, E. P.,** Inhibition of rat liver succinic dehydrogenase by selenium compounds, *J. Pharmacol. Exp. Ther.,* 108, 437, 1953.

96. **Klug, H. L., Moxon, A. L., Pedersen, D. F., and Potter, V. R.,** The *in vivo* inhibition of succinic dehydrogenase by selenium and its release by arsenic, *Arch. Biochem.,* 28, 253, 1950.

97. **McConnell, K. P. and Roth, D. M.,** Selenium-75 in rat intracellular liver fractions, *Biochim. Biophys. Acta,* 62, 503, 1962.

98. **Ames, B. N., McCann, J., and Yamasaki, E.,** Methods for detecting carcinogens and mutagens with the Salmonella/mammalian-microsome mutagenicity test, *Mutat. Res.,* 31, 347, 1975.

99. **Nakamuro, K., Yoshikawa, K., Sayato, Y., Kurata, H., Tonomura, M., and Tonomura, A.,** Studies on selenium-related compounds. V. Cytogenic effect and reactivity with DNA, *Mutat. Res.,* 40, 177, 1976.

100. **Jacobs, M. M., Matney, T. S., and Griffin, A. C.,** Inhibitory effects of selenium on the mutagenicity of 2-acetylaminofluorene (AAF) and AAF derivatives, *Cancer Lett.,* 2, 319, 1977.

101. **Lo, L. W., Koropatnik, J., and Stich, H. F.,** The mutagenicity and cytotoxicity of selenite, "activated" selenite and selenate for normal and DNA repair-deficient human fibroblasts, *Mutat. Res.,* 49, 305, 1978.

102. **Shamberger, R. J., Corlett, C. L., Beaman, K. D., and Kasten, B. L.,** Antioxidants reduce the mutagenic effect of malonaldehyde and B-proprolacactone. IX. Antioxidants and cancer, *Mutat. Res.,* 66, 349, 1979.

103. **Russell, G. R., Nader, C. J., and Patrick, E. J.,** Induction of DNA repair by some selenium compounds, *Cancer Lett.,* 10, 75, 1980.

104. **Ray, J. H. and Altenburg, L. C.,** Dependence of the sister-chromatid exchange-inducing abilities of inorganic selenium compounds on the valence state of selenium, *Mutat. Res.,* 78, 261, 1980.

105. **Rosin, M. P.,** Inhibition of spontaneous mutagenesis in yeast cultures by selenite, selenate, and selenide, *Cancer Lett.,* 13, 7, 1981.

106. **Martin, S. E., Adams, G. H., Schillaci, M., and Milner, J. A.,** Anti-mutagenic effect of selenium on acridine orange and 7,12-dimethylbenzanthracene in the Ames Salmonella/microsomal system, *Mutat. Res.,* 82, 41, 1981.

107. **Rosin, M. P. and Stich, H. F.,** Assessment of the use of the Salmonella mutagenesis assay to determine the influence of antioxidants in carcinogen-induced mutagenicity, *Int. J. Cancer,* 23, 722, 1979.

108. **Tkeshelashvili, L. K., Shearman, C. W., Zakour, R. A., Koplitz, R. M., and Loeb, L. A.,** Effects of arsenic, selenium and chromium on the fidelity of DNA synthesis, *Cancer Res.,* 40, 2455, 1980.

109. **Rasco, M. A., Jacobs, M. M., and Griffin, A. C.,** Effects of selenium on aryl hydrocarbon hydroxylase activity in cultured human lymphocytes, *Cancer Lett.,* 3, 295, 1977.

110. **Daoud, A. H. and Griffin, A. C.,** Effects of selenium and retinoic acid on the metabolism of N-acetyl-laminofluorene and N-hydroxyacetylaminofluorene, *Cancer Lett.,* 5, 231, 1978.

111. **Harbach, P. R. and Swenberg, J. A.,** Effects of selenium on 1,2-dimethylhydrazine metabolism and DNA alkylation, *Carcinogenesis,* 2, 575, 1981.

112. **Wortzman, M. S., Besbris, H. J., and Cohen, A. M.,** Effect of dietary selenium on the interaction between 2-acetylaminofluorene and rat liver DNA *in vivo, Cancer Res.,* 40, 2670, 1980.

113. **Bull, A. W., Burd, A. D., and Nigro, N. D.,** Effect of inhibitors of tumorigenesis on the formation of O^6-methylguanine in the colon of 1,2-dimethylhydrazine-treated rats, *Cancer Res.,* 41, 4938, 1981.
114. **Everett, G. A. and Holley, R. A.,** Effect of minerals on amino acid incorporation by a rat-liver preparation, *Biochim. Biophys. Acta,* 46, 390, 1961.
115. **Vernie, L. N., Bont, W. S., and Emmelot, P.,** Inhibition of *in vitro* amino acid incorporation by sodium selenite, *Biochemistry,* 13, 337, 1974.
116. **Vernie, L. N., Bont, W. S., Ginjaar, H. B., and Emmelot, P.,** Elongation factor 2 as the target of the reaction product between sodium selenite and glutathione (GSSeSG) in the inhibiting of amino acid incorporation *in vitro, Biochim. Biophys. Acta,* 414, 283, 1975.
117. **Vernie, L. N., Collard, J. G., Eker, A. P. M., DeWildt, A., and Wilders, I. T.,** *Biochem. J.,* 180, 213, 1979.
118. **Vernie, L. N., Homburg, C. J., and Bont, W. S.,** Inhibition of the growth of malignant mouse lymphoid cells by selenodiglutathione and selenodicysteine, *Cancer Lett.,* 14, 303, 1981.
119. **Yu, S.-Y. and Wang, L.-M.,** Differential effects of selenium on cyclic AMP metabolism in hepatoma cells and normal liver cells, *Biol. Trace Element Res.,* 5, 9, 1983.
120. **Spallholz, J. E., Martin, J. L., Gerlach, M. L., and Heinzerling, R. H.,** Immunological responses of mice fed diets supplemented with selenite selenium, *Proc. Soc. Exp. Biol. Med.,* 143, 685, 1973.
121. **Spallholz, J. E., Martin, J. L., Gerlach, M. L., and Heinzerling, R. H.,** Enhanced IgM and IgG titers in mice fed selenium, *Infect. Immun.,* 8, 841, 1973.
122. **Shakelford, J. and Martin, J.,** Antibody response of mature male mice after drinking water supplemented with selenium, *Proc. Am. Soc. Exp. Biol.,* 39, 339, 1980.
123. **Martin, J. L. and Spallholz, J. E.,** Selenium in the immune response, in *Proc. Symp. Selenium-Tellurium in the Environment,* Industrial Health Foundation, Pittsburgh, 1977, 204.
124. **Medina, D.,** unpublished observations.

Chapter 3

CHEMOPREVENTION OF CANCER

Raymond J. Shamberger

TABLE OF CONTENTS

I. INTRODUCTION

Because the mechanism of cancer formation has not been elucidated, one approach to the study of the problem has been to give animals another substance before or during carcinogen administration to see if carcinogenesis would be altered. By knowing something about the metabolism of the substance, certain insights into the mechanism of carcinogenesis could be made. In some cases dramatic chemopreventative effects have been observed. Some chemopreventatives prevent carcinogen formation, some prevent carcinogens from reacting or reaching critical target sites by enhancing detoxification mechanisms, and some are good "suppressing agents" and are effective when fed subsequent to carcinogen administration.

II. VITAMIN A

A. Effect on Animal Carcinogenesis

Of the many chemically diverse substances classified as vitamins, vitamin A and their retinoid derivatives are of the greatest current interest in terms of its possible role in the process of carcinogenesis. Vitamin A plays an important role in the control of growth, differentiation, and function of normal epithelial tissues. However, vitamin A deficiency results in hyperkeratosis of the skin, and also causes metablastic changes in the epithelia of the GI, respiratory, and urogenital tracts. Squamous type of epithelium lines the external part of the body and its orifices (mouth, pharynx, anus, and vagina). The gut, lower respiratory passages, gall bladder, and endocervix are lined by glandular epithelium. There have been several experiments suggesting that vitamin A deficiency is related to cancer of the skin, stomach, nasopharynx, lower respiratory tract, and endocervix.

When Rhino mice were fed a diet containing vitamin A in their feed, fewer 7,12-dimethylbenzanthracene (DMBA)-induced papillomas were observed than in mice fed a diet deficient in vitamin A.[1] When vitamin A was applied topically, the numbers of mouse skin papillomas induced by DMBA-croton oil was decreased.[2] The beneficial effect of vitamin A on the development of chemically induced papillomas may be due to its regulatory effect by controlling cellular differentiation of the epithelium.

Saffiotti et al. have reported that oral administration of vitamin A palmitate after benzopyrene treatment markedly inhibits not only squamous metaplasia, but also tracheal bronchiogenic carcinomas in hamsters.[3]

Vitamin A deficiency also affects the mucous of the urinary bladder, producing squamous cell metaplasia and a high incidence of cystitis, ureteritis, and pyelonephritis. When Sprague-Dawley rats were fed a diet deficient in vitamin A, there was an acceleration in the neoplastic response to N-[4-(5-nitro-2-furyl)-2-thiazolyl]-formamide (FANFT) which resulted in an earlier appearance of urinary bladder tumors and the development of ureteral and pelvic carcinomas.[4] In addition, a high incidence of squamous cell papilloma of the urinary bladder was observed in rats fed a vitamin A-free diet and given N-methyl-N-nitrosoguanidine (MNNG).[5] Chu and Malmgren have observed that a high dietary intake of vitamin A has inhibited the formation of tumors of the forestomach and the cervix in hamsters.[6] Merriman and Bertram have used cultured C3H/10T 1/2 clone 8 cells to study 3-methylcholanthrene (MCA)-induced neoplastic cell transformation. At concentrations that did affect cell survival, retinyl acetate was observed to inhibit MCA-induced cell transformation.[7]

The major advantage of retinoids is that they have the same anticarcinogenic effects of vitamin A, but are not as toxic as vitamin A. Several synthetic retinoids have been observed to prevent the development of skin cancer[8] of the respiratory tract,[9] cancer of the mammary gland,[10] and cancer of the urinary bladder.[11]

B. Effect on Mutagenesis

Vitamin A (retinol) has been shown to inhibit the mutagenic activity of aflatoxin B-1 when added to the Ames *Salmonella*/mammalian microsome test assay.[12] The inhibition was found to be dose dependent and was not caused by a direct toxic effect on the test bacteria. The same inhibitory effect was observed on the mutagenicity of an aminoazo dye, ortho-aminoazotoluene. The inhibition may be an effect on the mixed-function oxidases that convert ortho-aminoazotoluene to its ultimate mutagenic form. Retinol has inhibited sister chromatid exchange (SCE) frequencies and cell cycle delay in V79 cells induced by the indirect mutagen cyclophosphamide or aflatoxin B-1.[13]

C. Effects in Humans

The marked anticancer effect of vitamin A in animals has resulted in numerous epidemiological investigations, most of which have been case-control studies. These studies have indicated an inverse relationship between "vitamin A" intake and several different cancers. Most studies of vitamin A were based on the frequency of ingestion of a group of foods (e.g., green and yellow vegetables) which are rich in β-carotene (a provitamin that may be enzymatically converted to vitamin A in vivo) as well as a few foods such as whole milk and liver containing performed retinol (vitamin A).

Bjelke was one of the first investigators to report epidemiological data suggesting that vitamin A plays a protective role against cancer.[14] Using frequency of consumption data collected from a questionnaire mailed to a group of Norwegian men, Bjelke calculated a vitamin A index based on limited sources of the vitamin. After controlling for cigarette smoking, lower consumption values were observed for lung cancer cases than for controls. In another case-control study among Chinese females in Singapore, MacLennan et al. have observed an inverse association between consumption of green leafy vegetables rich in "vitamin A" and lung cancer.[15] Plasma levels of vitamin A were also lower in 28 patients with bronchial carcinoma than a small group of controls.[16]

Male cases of laryngeal cancer and controls have been studied by Graham et al.[17] After controlling for cigarette smoking and alcohol consumption, they found an inverse relationship (with a dose-response gradient) between cancer risk and both the vitamin A and C intake which were based on the frequency of consumption of selected foods. Similar results have been reported for vegetable consumption in general, but not for cruciferous vegetables in particular.

In a case-control study designed in a similar way as the one conducted on lung cancer, Mettlin et al. have observed a similar inverse association of their vitamin A consumption index with bladder cancer. In this study controlling was done for coffee consumption, smoking, and also occupational exposure.[18]

The frequencies of milk consumption, and of green and yellow vegetables (both sources of vitamin A and β-carotene) have been observed to be lower for esophageal cancer causes than for controls.[19] After controlling for cigarette smoking and alcohol consumption, a similar inverse association and a dose-response gradient for frequency of consumption of fruits and vegetables have been reported by Mettlin et al. in a study of male cases and controls.[20] Even though they also observed an inverse relationship for an index of vitamin A consumption based on selected foods, there was an even greater inverse relationship for an index of vitamin C consumption. These findings were consistent for populations in the Caspian littoral of Iran (a region of particularly high esophageal cancer incidence).[21]

In a case-control of gastric cancer in New York, Graham et al. have observed a higher consumption of uncooked vegetables (likely sources of β-carotene) by controls than by gastric cancer cases.[22] In a case control study in Hawaii, a similar inverse association with the consumption of raw vegetables was reported by Haenszel et al.[23]

In studies of cancer cases and controls in Norway and Minnesota, Bjelke has observed that milk and several vegetables have been eaten less frequently by colorectal cancer cases than by controls.[24] In addition, an index of vitamin A intake (which was highly correlated with vegetable consumption) showed the same inverse relationship.

Schuman et al. have related foods rich in vitamin A (e.g., liver) and β-carotene (e.g., carrots) to prostate cancer and found that the vitamin A rich foods were consumed less frequently by cases than by controls.[25]

Women with abnormal cytology were matched with normal control subjects of age, parity, ethnicity, and socioeconomic class.[26] The women participated in a blind case-control study focused on a nutritional survey to relate nutrition to cervical dysplasia. The nutritional survey revealed statistically significant differences for vitamins A and C and beta carotene. Serum retinol binding protein was absent or minimally detectable and inversely related to the severity of the dysplasia.

In three recent reports based on data from group studies in the U.S. and England, the investigators observed that there was an inverse relationship between serum levels of vitamin A and the subsequent risk of cancer in general.[27] There is also a known correlation between retinol and cholesterol concentrations and cancer mortality which may reflect a relationship between their carrier proteins-retinol binding protein and low-density lipoprotein. A study of serum cholesterol, serum retinol, and serum carotene concentrations in relation to the incidence of cancer may help elucidate the cholesterol-cancer relationships.

D. Mechanisms of Action

There are several possible mechanisms for the anticancer effect of vitamin A.

1. Mixed-Function Oxidase

Several vitamin A compounds and analogs are able to inhibit the in vitro microsomal mixed-function oxidases that metabolize carcinogenic polycyclic aromatic hydrocarbons (PAH).[28] This may be an important mechanism involved in the protection of vitamin A-induced squamous metaplasia. In addition, Carruthers has reported that intraperitoneally administered 3,4-benzopyrene and methylcholanthrene results in a marked reduction of hepatic vitamin A in rats.[29] These results suggest that the depletion of vitamin A by PAH may be important in the development of carcinogenesis.

2. DNA Binding

The binding of carcinogens to DNA has been suggested to be a critical step in the carcinogenic process. For PHA the binding to DNA may be a measure of the carcinogenic potential towards any given tissue. Vitamin A deficiency has been reported to increase the binding of benzopyrene to the DNA of hamster tracheal cells.[30]

3. Effect on Metabolic Activation

Huang et al. have reported that retinol neither affects the frequency of SCEs nor cell cycle delay in Chinese hamster V79 cell with or without metabolic activation by S-9 mix, but inhibits the SCE frequencies and cell cycle delay in V79 cells induced by the indirect mutagen cyclophosphamide or aflatoxin B-1.[13] The results suggest that retinol itself may not alter genetic materials but may exert its effect by inhibiting the metabolic activation of an indirect mutagen or carcinogen.

4. Inhibition of Ornithine Decarboxylase

Application of the potent tumor promoter 12-*O*-tetradecanoylphorbol-13-acetate (TPA) to mouse skin resulted in a 200-fold increase in epidermal ornithine decarboxylase activity.[31] This phenotypic change may be essential for skin tumor promotion. Verma et al. found that vitamin A and its analogs applied to the skin decrease both the TPA-induced ornithine decarboxylase activity and the formation of skin papillomas. Retinoic acid administered systematically also inhibited ornithine decarboxylase activity. In another type of experiment hairless mice were irradiated once daily with UV light.[32] In the experimental group topical retinoic acid was applied immediately after each irradiation. After 52 weeks, the groups treated with retinoic acid generally had fewer mice with tumors, fewer tumors per mouse, smaller tumor diameters and slower growing tumors than did the irradiated control groups. Retinoic acid also decreased the activity of UV light induced epidermal ornithine decarboxylase activity.

5. Effect on Lysosomal Membranes

Vitamin A and the polyene antibiotics are known to labilize lysosomal membranes. Retinoids also labilize lysosomal membranes which are able to act on preneoplastic and neoplastic cells alike. When Lotan and Nicolson incubated retinoids with tumor cells in vitro, the viability of the treated cells was not altered.[33] Direct measurements of free lysosomal enzymes in the cytoplasms of the melanoma cell lines S91 and B16, which are known to be sensitive to inhibition of growth by retinoic acid, did not demonstrate any increased release or change in the ratio of free/bound acid phosphatase or arylsulfatase. Therefore, lysosomal membrane labilization may not be an important mechanism in the chemopreventative effect of vitamin A on tumors.

6. Effect of Immune Functions

Due to their presence in certain organs, tissues, and body fluids other than blood plasma, such as lymph, vitamin A and β-carotene may influence the growth and maturation of cells that regulate various immune functions, such as thymus function, thereby decreasing tumor growth. Stimulation of the immune response by vitamin A compounds is adequately documented and reviewed by Hill and Grubbs.[14] Vitamin A administration can cause greater humoral responses to foreign proteins and to heterologous red blood cells, accelerated graft rejection, increased cellularity of lymph nodes draining injection depots, and increased resistance to infection with a variety of bacterial pathogens.

III. ASCORBIC ACID

A. Effect on Animal Carcinogenesis

When ascorbic acid (0.2%) was applied at the same time as croton oil to mouse skin which had been previously treated with DMBA, the total number of mouse skin papillomas was decreased.[34] However, vitamin C did not seem to reduce tumors as much as selenium and vitamin E in the same experiments. Slaga et al. have also observed a reduction in the number of skin tumors induced by DMBA-phorbol carcinogenesis in mice and treated similarly with ascorbic acid.[35] Sadek and Abdelmegid have observed that toads receiving 10 mg/kg/day of ascorbic acid for 8 weeks have considerably less DMBA-induced skin tumors than the controls.[36] Ascorbic acid has also decreased the incidence of large malignant skin tumors and other lesions in hairless mice subjected to UV light over a period of 15 weeks.[37]

Reddy et al. have studied the effect of dietary sodium ascorbate on colon carcinogenesis induced by 1,2-dimethylhydrazine (DMH) or *N*-methyl-*N*-nitrosourea (MNU) in female F344 rats.[38] The incidence of colon and kidney tumors was lower in those animals fed 0.25 or

1.0% sodium ascorbate and treated with a single dose of DMH than the incidence in the animals fed a control diet without sodium ascorbate. However, the tumor frequencies did not differ between the sodium ascorbate and control diet-fed animals which were treated with multiple doses of DMH or MNU. When 1.2% ascorbic acid was added to the low-fat diet of Sprague-Dawley rats DMH-induced colon tumor incidence was also inhibited.[39] In contrast, no significant effect dietary ascorbic acid was observed on the development of DMH-induced colon tumor in female CF, mice fed a low-fat diet.[40]

B. Effect on Mutagenesis

Wirth et al. have studied the mutagenicity of phenacetin and acetaminophen and their respective N-hydroxylated metabolites in *Salmonella typhimurium*.[41] Ascorbic acid reduced the mutation frequency (80 to 90%) of *N*-hydroxyphenacetin in both TA 98 and TA 100 as well as *p*-nitrosophenetole in TA 100, but slightly increased the mutation frequency of *N*-hydroxy-2-aminofluorene and 2-nitrosofluorene in both TA 98 and TA 100.

Shamberger et al. have reported that ascorbic acid as well as other antioxidants including butylated hydroxytoluene, vitamin E, and selenium, prevented mutagenesis caused by the direct acting mutagens, malonaldehyde or beta-propiolactone, when several varieties of *Salmonella* strains were tested with the antioxidants.[42] Malonaldehyde affected the strains with tendency to mutate through frame shift mutagenesis while beta-propiolactone acted on tester strains which are known to mutate through both frame shift mutagenesis and base pair substitution.

N-nitrodimethylamine, *N*-nitrodiethylamine, *N*-nitromorpholine, and their *N*-nitroso analogs, *N*-nitrosodimethylamine and *N*-nitrosomorpholine, were tested in *Salmonella typhimurium* TA 100 and TA 1530.[43] All of the compounds except *N*-nitrodiethylamine were mutagenic. The mutagens required activation by the presence of postmitochondrial supernatant from the liver of Aroclor-treated rats, oxygen, and the reduced nicotinamide adenine dinucleotide phosphate-generating system. Addition of ascorbic acid and disulfiram to the assays were effective in inhibiting mutagenesis by all nitro and nitroso compounds. The mutagenesis induced by MNNG and dimethyl-nitrosamine (DMN) in Salmonella TA 1530 was also inhibited by ascorbate.[44] The reduction of MNNG-induced mutagenesis resulted from a reaction that took place between ascorbic acid and MNNG that consumed MNNG. This reaction was greatly enhanced by catalytic amounts of Cu^{2+} and Fe^{3+}. No direct reaction was detected between DMN and ascorbate. Mutagenesis by *N*-methyl-*N*-nitrosourea was not reduced by ascorbate. Using several *Salmonella* tester strains, nitrosated spermidine was reported to be mutagenic but the mutagenesis was blocked by ascorbic acid.[45] A mutagenic response was obtained in the *Salmonella* test by preincubating sodium nitrite and cimetidine in human gastric juice from untreated individuals.[46] Ascorbic acid was efficient in blocking the formation of mutagenic nitro-derivatives. Ascorbic acid was found to decrease the amino-pyrine/nitrite-induced mutation frequency in *E. coli*, but there was no increase or decrease in the dimethylnitrosamine-induced mutation frequency observed in the presence of ascorbic acid.[47]

Lin et al. have tested the mutagenicity of several brands of soybean sauce in vitro.[48] When treated with nitrite at 2000 ppm, soybean sauce produced a mutagenic substance when several strains of *Salmonella typhimurium* and the mammalian microsome mutagenicity test was used. All 21 different brands of soybean sauce were mutagenic. The greatest amount of mutagenic material was formed when the nitrite level was 2000 ppm and the pH was 3. Ascorbic acid seemed to prevent the formation of mutagenic products in nitrite-treated soybean sauce. Soy sauce is one of the extracts of food materials used extensively by Chinese people in food preparation. The mutagenesis observed with soy sauce and nitrite may be related to the epidemiologic observations that carcinomas of the liver and stomach are the most common malignant tumor among Chinese people living in Taiwan.

Lee et al. homogenized seven species of raw fish commonly eaten by Koreans and nitrosated them in vitro under simulated gastric condition with and without ascorbic acid and then tested them with the *Salmonella* Ames test.[49] Ascorbic acid, in an amount three- to fivefold equimolar to the nitrate used, almost completely blocked the formation of the mutagenic principle.

Munkres has reported that ferrous ions were both highly lethal and mutagenic to germinated condidia of *Neurospora crassa*.[50] In these experiments, treatment with 0.2 mM ferrous ions was 14- and 50-fold more mutagenic than UV irradiation or X-rays, respectively. Ascorbic acid at 2 mM was not lethal, but reduced both the lethality and mutagenicity of ferrous ions. The residual lethality of ferrous ions was also completely inhibited by bovine superoxide dismutase (SOD). The protection of these experiments by ascorbic acid and SOD indicates that the superoxide radicals generated by the oxidation of Fe^{2+} could be directly or indirectly mutagenic and lethal.

C. Effects in Humans

Schlegal et al. observed that vitamin C reduces uroepithelial carcinoma in mice and also suggested that a similar mechanism exists in humans.[51] The tryptophan metabolite 3-hydroxyanthranilic acid (3-HOA) is thought to be stabilized by ascorbic acid, thereby preventing its carcinogenicity when 3-HOA is implanted in the bladder. In one study ten bladder tumor patients were found to excrete urine with higher concentrations of both 3-HOA and 3-HOK (3-hydroxykynurenine) than urine excreted by healthy controls.[52] Price et al. have reported elevated levels of various chemiluminescent tryptophane metabolites, such as 3-HOA in the urine of patients with bladder cancer.[53] In addition, about 50% of their tumor patients excreted high concentrations of 3-HOK in their urine. In contrast, Benassi et al. found lower than normal instead of elevated concentrations of 3-HOA in the urine of bladder tumor patients.[54] Pipkin et al. have attempted to explain the difference in the three experiments[55] by checking the stability of 3-HOA in the urine of normal patients and patients with bladder tumors. About 90% of the 3-HOA was recovered in normal patients after an incubation at 37°C for 6 hr. However, in patients with bladder tumor, recoveries below 50% were found. The difference in the recovery level of normal and bladder patients suggests a difference in the redox environment of the urine. When ascorbic acid was given orally, urinary 3-HOA was stabilized. This experiment suggests that ascorbic acid may be directly or indirectly related to the stabilizing agent.

Fecal extracts from many normal individuals contain mutagens that can be detected with the *Salmonella* tester strains.[56] The occurrence of mutagenically active donors was greater in populations on Western diets than it is in populations on Black South African, vegetarian, or Japanese diets. The differences between these population groups suggest that mutagen levels are affected by the type of diet. Bruce et al. have also observed that the levels of fecal mutagens are reduced by >60% in all donors when their normal diets were supplemented with 4 g of vitamin C per day.

Dion et al. have tested the effect of supplemental ascorbic acid and alpha-tocopherol on fecal mutagenicity in two separate studies involving 20 healthy donors, aged 22 to 55 years.[57] The mutagens were extracted from individuals's frozen feces samples with dichloromethane and assayed with *Salmonella typhimurium* tester strain TA 100 without microsomal activation. In the first study, with only a single donor on a controlled diet, the fecal mutagenicity decreased on treatment to 21% of the control. In the second study, when 19 donors were on free-choice diets, the mutagenicity was reduced on treatment ($p < 0.01$) to 26% of control. Both vitamins E and C diet may have a role in lowering the body's exposure to endogenously produced mutagens. The mechanism responsible for the reduction of mutagenesis by ascorbic acid and alpha-tocopherol is unknown. Substantial variability in the mutagen levels was observed on a day-to-day basis. Much of this variability is likely a result of daily physiological changes which might be associated with transit time as the mutagen seems to be a product of colonic bacterial metabolism. The concentration of mutagens in the feces increases with the duration of fecal anaerobic incubation at 37°C. However, there is no direct evidence that fecal mutagenicity is related to large bowel cancer or any other human disease. Nor is it known whether a deficiency in antioxidants could lead to an increased malignancy risk.

Volatile mutagens were produced from normal human and animal feces when incubated with sodium nitrite in saline at 37°C for 48 hr.[58] The mutagens were detected using the Ames *Salmonella* tester strain TA 1535 without microsomal activation. Sodium ascorbate and alpha-tocopherol each reduced the mutagenicity by about 30%.

Daily doses of 1.0 g ascorbic acid were given to a group of 35 coal-tar workers occupationally exposed to PAHs and benzene during the processing of coal tar.[59] Genetic analysis of peripheral blood lymphocytes showed a significant drop in the frequency of aberrant cells.

D. Mechanism of Action

Vitamin C prevents carcinogenesis through a variety of mechanisms. Vitamin C may prevent tumorigenesis through its antioxidant action. The antioxidant effect may be through the prevention of a perioxidation which might cause free radical damage to various cellular components.

Mutations may arise either directly by free radical-mediated oxidative deamination of DNA bases or may arise indirectly through the reactions of free radicals with DNA polymerases which may result in a decrease of the fidelity of replication or repair. The partial inactivation of the error-prone *E. coli* repair polymerase in vitro by superoxide radicals, generated by xanthine oxidase for example, leads to a marked increase in the incorporation of the wrong nucleotide, dCTP, relative to the correct one, TTP, with template of synthetic poly (dAT). In addition, the inactivation and the alteration of the synthetic fidelity are known to be prevented by bovine SOD.

Shoyab has studied the effect of vitamin C on the binding of tritiated DMBA to murine epidermal cells DNA in culture. Both vitamin C and its salt significantly reduced the binding of DMBA to DNA.[60] In this way, initiation of the cancer process may be reduced.

Ascorbic acid blocks the in vitro formation of carcinogenic *N*-nitroso compounds by the reaction between nitrous acid and oxytetracycline, morpholine, piperazine, *N*-methylaniline, methylurea, and dimethylamine. The extent of blocking depends on the compounds nitrosated and the experimental conditions.[61]

Historically, salt or saltpeter has been used for the preservation of food, and the crude salt frequently contains large amounts of nitrate. A high nitrate content is also present in fertilizers and the nitrate is incorporated into foodstuffs, especially into such vegetables as spinach and carrots. In addition, nitrate is frequently added to pork to prevent oxidation of its high content of unsaturated fat which leads to peroxidation and the formation of rancid products. Nitrite also was extensively added to meat to make it look redder and therefore more desirable for purchase, but this practice has been generally curtailed. In an experiment with boiled potatoes, added nitrite levels were blocked by the addition of ascorbate.[62] The species formed from nitrous acid, which are responsible for the oxidation of ascorbic acid, are also the same species which cause nitrosation of secondary amines.[63] Between pH 1.5 and 5.0, the nitrosation of secondary amines in the absence of oxygen and the presence of ascorbic acid can be summarized by two competitive parallel second-order reactions.

$$\text{Amine} + N_2O_3 \xrightarrow{k_1} \text{nitrosamine} + NO_2^- + H^+$$

$$\text{Ascorbate} + N_2O_3 \xrightarrow{k_2} \text{dehydroascorbate} + 2\,NO + 2\,H_2O$$

If $k_2 >> k_1$, then Reaction 2 is mostly complete before 1 starts. Large doses of vitamin C have protected rats from liver tumors induced by aminopyrine and sodium nitrate.[64] The mechanism may result, in part, from blockage of in vivo nitrosation, which causes dimethylnitrosoamine formation. However, ascorbic acid did not completely protect against lung and kidney tumor formation. Perhaps the concentration of ascorbic acid did not reach sufficient levels in the kidney and the lung to block in vivo nitrosation.

Werner et al. have studied the effect of ascorbic acid on small intestine cancer included by *N*-ethyl-*N*-nitro-*N*-nitrosoguanidine (ENNG) in rats.[65] The induction of tumors was not affected by amounts as large as 2 to 3% of sodium ascorbate in the diet, but the depth of tumor infiltration was restricted by sodium ascorbate.

IV. CHOLINE

A. Effect on Animal Carcinogenesis

Rats fed choline-deficient diets developed liver tumors which have been linked to the presence of aflatoxin. However, when the diet was supplemented with choline, inhibition of tumor incidence was observed.[66] Lombardi and Skinozuka have reported that a choline-deficient diet enhances the induction in rats of liver tumors induced by 2-acetylaminofluorene.[67]

Shinozuka et al. have observed a 60 to 70% incidence of hepatocellular carcinomas after 4 to 5 months when azaserine, which is ordinarily a weak inducer of liver tumors, was administered to choline-deficient rats.[68] When ethionine was fed to rats maintained on a methionine-choline-deficient diet, liver tumor induction was also markedly increased.

Feeding a choline-deficient diet to rats strongly promotes the progression of liver cells initiated by a liver carcinogen diethylnitrosamine, to foci of gamma-glutamyltranspeptidase (GGTP)-positive hepatocytes. Takahashi et al. have studied whether or not a choline-deficient diet might promote the progression of GGTP-positive foci to hepatomas.[69] During the first 7 weeks after initiation, the number and size of the foci increased. The rats were then divided into two groups: one group received a choline-supplemented diet and one group remained on the choline-deficient diet. At 12 to 16 weeks the number and size of foci began to decline in rats fed the choline-supplemented diet. However, in contrast, there was a progressive increase in the size with coalescence of the foci, as well as development of the neoplastic nodules in the choline-deficient group.

Shinozuka et al. have initiated rats with diethylnitrosamine (DEN) and fed them choline-supplemented or choline-deficient diets in which the degree of saturation of the fat was obtained by using hydrogenated vegetable oil and corn oil, either alone or in combination.[70] The number of GGTP-positive foci developed after 7 weeks of promotion by the choline-deficient diet with predominantly corn oil was 2.6 times greater than that by the diet containing a mixture of hydrogenated vegetable oil and corn oil. In addition, the number of hepatocytes in the mitotic stage in the livers of rats fed the choline-deficient diet was three times higher than those in the livers of rats fed the choline-deficient diets with a mixture of hydrogenated vegetable oil and corn oil or with predominantly hydrogenated vegetable oil.

Ghoshal and Farber have observed that male Fischer-F344 rats on a choline- and a methionine-deficient (CMD) diet developed GGTP-positive foci which later developed into hepatocellular carcinoma.[71] The hepato-carcinogenic effect of CMB may not be due to contaminating aflatoxins (less than 5 ppb), volatile nitrosamines (1 to 2 ppb or less), and no measurable nitrites and nitrates. Similar results have been reported by Poirier et al.[72] In view of these initial observations of the induction of liver cancer by the CMD diet without carcinogens, more research is needed in this promising area.

B. Effect on Mutagenesis in Bacteria

Reddy et al. have used *Salmonella* mutagenesis assays to evaluate the mutagenicity of several chemical carcinogens using liver S-9 fractions from rats fed either a choline-supplemented or choline-deficient (CD) diet.[73] The liver S-9 fraction from CD diet-fed rats was observed to have a significantly decreased ability to activate 2-acetylaminofluorene, 2-aminoanthracene fluorene (HO-*N*-2-AAF), and dimethylnitrosamine. The same CD liver S-9 fraction was much less effective in deactivating MNNG, but not MNU.

C. Effects in Humans

The human requirement of choline is related to some extent to methionine availability, as well as the folacin and B-12 content of the diet. Because of the interaction of choline with these and other dietary components, neither the human requirement for choline nor its effect on human cancer is known.

D. Mechanism of Action

The results of the study by Reddy et al. suggest that feeding a choline-deficient diet impairs the activation of procarcinogens to reactivate species which are mutagenic.[73] A similar conclusion has been previously made by Rogers, who studied the effects of feeding a multiple lipotrope-deficient diet on the liver activation of several chemical carcinogens.[74] However, the studies of Reddy et al. have also suggested that in the case of proximate carcinogens, such as HO-*N*-2-AAF and of direct carcinogens, such as MNU and MNNG, the choline-deficient diet had no effect, or may actually enhance their carcinogenicity by blocking their conversion to inactive metabolites.[73]

Giambarresi et al. have reported that feeding a choline-deficient diet to rats causes a loss of prelabel DNA, and therefore of liver cells, accompanied by a stimulation of liver cell proliferation.[75] The authors conclude that feeding a choline-deficient diet starts a low level of liver-cell regeneration. It may be in this way that a choline-deficient diet acts as a promotor of the further evolution to neoplasia of carcinogen-initiated liver cells. Because of the enhancement of carcinogenesis by unsaturated fat,[50] lipid peroxidation may also be an important factor.

V. VITAMIN E

A. Effect on Animal Carcinogenesis

The number of mouse skin papillomas induced by DMBA-croton oil was reduced by an alpha-tocopherol and vitamin C. Vitamin E may prevent mouse skin tumorigenesis through its antioxidant effects.[76] Vitamin E was more effective than vitamin C in these experiments. Jaffe has reported that a diet containing wheat germ oil decreased the number of a mixed group of tumors resulting from the i.p. injection of 3-methylcholanthrene (MCA).[77] In contrast, wheat germ oil added to the diet of mice apparently did not influence tumor development induced by painting the skin with benzopyrene.[78] Alpha-tocopherol and a number of phenolic antioxidants did not reduce the induction of s.c. sarcoma in the mouse by injection of 3,4,10-dibenzpyrene.[79] In the latter two experiments, the potential anticarcinogenic effect of vitamin E may have been overwhelmed by large amounts of carcinogen. Haber and Wissler added vitamin E to the diet of mice and observed a decreased incidence of s.c. sarcomas included by MCA.[80] Harman has studied the potential role of free radical reactions in aging and a number of pathological processes which include neoplasia.[81] In his experiment, rats fed a diet containing large amounts of vitamin E had significantly less mammary tumors (8/20) induced by DMBA than did the controls (14/19). Ip, however, reported vitamin E had no effect on DMBA carcinogenesis in rats.[82] Narayan has observed that vitamin E-deficient diets did not accelerate the induction of rat liver tumors induced by *N*-2-fluorenylacetamide in rats.[83]

Shklar has observed that Syrian golden hamsters given oral vitamin had had both fewer and smaller buccal pouch cancers induced by DMBA.[84] Microscopic examination showed that there was less invasion of underlying tissues and less surface necrosis. Pauling et al. have found that a vitamin E intake had no effect on the incidence of squamous cell carcinoma in hairless mice irradiated with UV light.[85] However, vitamin E has been shown to induce morphological differentiation and also increases the effect of ionizing radiation on neuroblastoma cells in culture.[86]

Cook and McNamara have reported that dietary vitamin E fed at 600 mg/kg has reduced the number of colonic tumors induced by dimethylhydrazine (DMH).[87] In contrast, Toth and Patil have observed that 4% dietary vitamin E increased the incidence of tumors in the duodenum, cecum, colon, rectum, and the anus.[88] In the latter experiment the amount of vitamin E in the diet was substantially greater and the animals were sacrificed later in the experiment.

B. Effect on Mutagenesis

Shamberger et al. has observed that vitamin E decreases the in vitro DMBA-induced mutagenesis of human lymphocytes.[89] Vitamin E has also reduced the percentage of benzopyrene included chromosomal abberations in both cultured Chinese hamster cells and Chinese hamster ovary cells.[90] The effect of 400 mg of both supplemental ascorbic acid and alpha-tocopherol and fecal mutagenicity has been examined in two studies involving 20 healthy human donors aged 22 to 55 years by Dion et al.[91] In one study the mutagen was extracted with dichloromethane from individual frozen feces samples and assayed with *Salmonella typhimurium* tester strains TA 100 without microsomal activation. In two different experiments, fecal mutagenicity decreased on treatment to 2 and 26% of the control. In the other study ascorbic acid and alpha-tocopherol were added directly to feces and there were no changes in mutagenicity. Apparently, antioxidants in the diet may have a role in reducing the body's exposure to endogenously produced mutagens. Kalina et al. have observed that vitamin E had an antimutagenic effect on the frequency of gene mutations induced by *N*-nitroso-*N*-methylurea in Salmonella.[92]

C. Effects in Humans

Although mammals in vitamin E-deficient diets may show a wide spectrum of pathologic conditions, there is no evidence to indicate that man is susceptible to vitamin E deficiency when he consumes an average American diet. The average intakes of alpha-tocopherol by adults is about 15 mg/day, but the variation could be quite large. Individuals fed diets high in protein and low in plant fat would consume less than 10 mg/day, whereas individuals consuming diets high in polyunsaturated oils might take in 60 mg/day.

The recommended daily intake of vitamin E is 10 mg/day in adult males, and 8 mg in adult females. In children ages 1 to 10, 5 to 7 mg are recommended.

A greater requirement for vitamin E is thought to be needed for individuals consuming a high polyunsaturated fat diet, especially when the fats are oxidized or contain large amounts of fish oil. Fish oils have a high peroxidative potential and low levels of tocopherol.

Weisburger et al. have observed a greater incidence of stomach cancer in populations consuming low levels of vitamin E and certain other selected micro nutrients.[93] They suggested that ingested nitrites or nitrates may be metabolized to carcinogenic alkylnitrosourea compounds in the stomach of rats. Vitamin E, as well as vitamin C, can inhibit the nitrosations that result in gastric nitrosourea formation. Because vitamin E is a lipid-soluble vitamin and vitamin C is a water-soluble vitamin, they may complement each other as carcinogenic inhibitors. Weisburger et al. suggest that gastric cancer is a preventable disease and its incidence may be reduced by serving foods containing vitamin C and perhaps vitamin E with every meal.[93]

Noncancerous lumpy breast tissue is present in about 20% of American women. This disorder has been termed either fibrocystic breast disease, mammary dysplasia, or fibrous mastopathy. Abrams[94] and London et al.[95] have reported that vitamin E produced moderate to complete relief of symptoms in several women with palpable softening of the breasts, and with reduction in cyst size.

D. Mechanisms of Action

The anticarcinogenic effects of vitamin E appear to be obtained mainly at higher non-physiological concentrations. The vitamin probably acts through its effect of reducing lipid peroxidation. There is also a possibility that vitamin E may have an effect on the metabolism of selenium, another antioxidant. Ip and Horvath have observed that vitamin E potentiated the inhibitory effect of selenium on DMBA mammary carcinogenesis only during the promotion or proliferative phase.[96]

Another possible anticarcinogenic effect of vitamin E might be on mitosis and DNA production, Konings and Trieling have observed an enhanced inhibition of ^{3}H-thymidine incorporation into DNA of vitamin E-depleted lymphosarcoma cells.[97]

Aryl hydrocarbon hydroxylase (AHH) is a microsomal membrane bound mixed-function oxidase which metabolizes PAHs. Maintenance of rats on a vitamin E-deficient diet for 12 weeks has resulted in a significant reduction of hepatic AHH activity.[98]

VI. PHENOLIC ANTIOXIDANTS (BHA, BHT)

A. Effect on Animal Carcinogenesis

BHA has been proven to be an important inhibitor of carcinogenesis and has been extensively studied for its ability to inhibit carcinogen-induced neoplasia. Tables 1 and 2 lists experiments in which BHA and BHT have been shown to have inhibitory effects. In these studies, BHA was administered before and/or during exposure to the carcinogen. If BHT is administered during the promotion or proliferative phase of urethane, 3-methylcholanthrene, or diethylnitrosamine induced carcinogenesis, the number of tumors increase.[99] In this regard BHT has been regarded as having tumor-promoting properties.

B. Effect on Mutagenesis in Bacteria

BHA has been shown to inhibit host-mediated mutagenesis due to exposure to hycanthone, metrifonate, praziguantel, and metronidazole.[100] BHT has decreased the *Salmonella* reversion rate stimulated by mutagens such as benzopyrene and its derivatives,[101] aminoanthracene, homidium and proflavine,[102] malondialdehyde, and beta-propiolactone.[103] BHT has also decreased the amount of DMBA-induced chromosomal damage when peripheral human white blood cells were incubated with hepatocarcinogens.[104]

C. Effects in Humans

Butylated hydroxytoluene (BHT) and butylated hydroxyanisole (BHA) have been added to foods, cosmetics, drugs, and animal feeds to prevent oxygen-induced lipid peroxidation. Nearly 9 million lb of BHA and BHT were produced for use in U.S. foods in 1976. The estimated daily intake of BHT is 0.5 to 1.5 mg/kg body weight in persons over 2 years old. The level of BHA and BHT used in food are regarded as safe in regard to toxicity. There are no epidemiological studies concerning the effects of BHT and BHA on human health.

D. Mechanism of Action

Studies of the mechanism by which BHA inhibits chemically induced carcinogenesis have shown that this phenolic compound produces a multiple enzyme response that may cause a greater rate of detoxification.[106] Increases in both glutathione S-transferase and tissue glutathione levels have been observed in mice that have been fed BHA for 1 to 2 weeks in carcinogen inhibition experiments.[107] Glutathione S-transferase is an important enzyme for detoxifying chemical carcinogens. Glutathione S-transferases promote the conjugation of many electrophilic and usually lipophilic compounds with glutathione.

Uridine diphosphate (UDP)-glucuronyl transferase which is another important conjugated enzyme in the detoxification system, is also increased.[108] The feeding of BHA has also been

Table 1
INHIBITION OF CARCINOGEN-INDUCED NEOPLASIA BY BHA

Carcinogen inhibited	Species	Site of neoplasm
Benzo[a]pyrene	Mouse	Lung
Benzo[a]pyrene	Mouse	Forestomach
Benzo[a]pyrene-7-8-dehydrodiol	Mouse	Forestomach, lung, and lymphoid tissue
7,12-Dimethylbenz[a]anthracene	Mouse	Lung
7,12-Dimethylbenz[a]anthracene	Mouse	Forestomach
7,12-Dimethylbenz[a]anthracene	Mouse	Skin
7,12-Dimethylbenz[a]anthracene	Rat	Breast
7-Hydroxymethyl-12-methylbenz[a]anthracene	Mouse	Lung
Dibenz[a,h]anthracene	Mouse	Lung
Nitrosodiethylamine	Mouse	Lung
4-Nitroquinoline-N-oxide	Mouse	Lung
Uracil mustard	Mouse	Lung
Urethan	Mouse	Lung
Methylazoxymethanol acetate	Mouse	Large intestine
trans-5-Amino-3-[2-(5-nitro-2-furyl)vinyl]-1,2,4-oxadiazole	Mouse	Forestomach, lung, and lymphoid tissue

From *Diet, Nutrition and Cancer*, National Academy Press, Washington, D.C., 1982, 15-1. With permission.

Table 2
TUMOR PROPHYLACTIC EFFECTS OF BHT

Carcinogen inhibited	Species	Site of neoplasm
DMBA	Mouse	Forestomach
Diethylnitrosamine (DEN)	Mouse	Forestomach
DMBA	Mouse	Skin
Urethane	Mouse	Lung
DMBA	Rat	Mammary gland
2-acetylaminofluorene (AAF)	Rat	Hepatoma
3-methyl-4-dimethyl-aminoazobenzene	Rat	Hepatoma
Axoxymethane (AOM)	Rat	Intestine

reported to increase epoxide hydrolase activity[109] and also altered the microsomal monoxygenase system.[110]

BHA or BHT may also affect the binding of carcinogens to DNA. Liver microsomes from eight strains of mice fed either a diet containing BHA, or a single dose of BHA, had a reduced capacity (24 to 60% of control values) for generating benzopyrene metabolites that could bind to exogenous calf thymus DNA.[111] The reduction of binding of the carcinogen paralleled protection against the formation of pulmonary adenomas by benzopyrene.

VII. SELENIUM

A. Effect on Animal Carcinogenesis

Selenium has been extensively studied as an inhibitor of animal carcinogenesis. Sodium selenide applied concomitantly along with croton oil has greatly reduced the number of mice with skin tumors after the mice had been initiated with DMBA.[113] Mice supplemented with sodium selenite also markedly reduced papillomas induced by the carcinogens, DMBA-croton oil. A similar reduction with dietary selenium was observed with benzopyrene.[113]

Inclusion of selenium in the diet has also reduced the number of liver tumors induced by 3-methyl-4-dimethylaminoazobenzene (DAB),[114,115] diethylnitrosamine,[116] and acetylaminofluorene.[117]

Dietary selenium has its greatest carcinogenesis-inhibiting effect on colon and breast cancer. Sodium selenite in the drinking water has markedly inhibited 1,2-dimethylhydrazine (DMH)-induced carcinogenesis.[118,119] Spontaneous breast cancer induced by Bittner milk virus and DMBA-induced breast cancer have also been markedly reduced by dietary selenium.[120-122] Dietary selenium has also reduced the number of tumors induced by weekly injections of bis(2-oxopropyl) nitrosamine (BOP).[123]

B. Effect on Mutagenesis

Selenium has been shown to have both mutagenic and also antimutagenic effects in bacterial and mammalian cells. The effect seems to be dependent on the concentration of selenium. Physiological concentrations of selenium seem to inhibit the mutagenic process whereas concentrations of selenium ten or more times of that found in normal cells enhance the mutagenic process perhaps due to a genotoxic effect.

Using the *Salmonella typhimurium* test system of Ames, added selenium has reduced the mutagenicity of 2-acetyl aminofluorene (AAF), *N*-hydroxy-2-acetylaminofluorene (*N*-OH-AAF), *N*-hydroxyaminofluorene (*N*-OH-AF), DMBA, sodium nitrite,[103,124-126] acridine orange, malonaldehyde, and beta-propiolactone. Selenium has also been shown to reduce the mutagenicity of known mutagens in human lymphocyte cultures.[104,127]

C. Effects in Humans

There have been two major studies of epidemiological relationships between environmental selenium and cancer in two different populations. Even though both relationships were inverse to selenium bioavailability and paralleled the animal results, this does not necessarily mean there is a real relationship. In one study a series of epidemiological studies has related selenium bioavailability in various cities and states to the human cancer mortality in the cities and states of the U.S.[128,129] Statistically significant differences were observed for the age-specific cancer death rates among states with high, medium, and low selenium levels. The death rates for several types of cancer were larger in males than in females in the states with high selenium levels. The greater difference between males and females may have been due to sex difference or because males are heavier smokers and are more likely to be exposed to industrial carcinogens. In the states with high selenium forage levels, there was significantly lower mortality in both males and females from several types of cancer, particularly the environmental problem indicators, such as the GI and urogenital types of cancer. A similar observation was observed in 17 paired large cities and 20 paired small cities.[128] Alberta and Saskatchewan had a higher selenium bioavailability and also had a lower human cancer death rate than the other Canadian provinces.

In a second major epidemiological study, Schrauzer et al. have correlated the age-corrected mortality from cancer at 17 major body sites with the apparent dietary selenium intakes estimated from food-consumption data in 27 countries.[130] There were also significant inverse correlations found for cancers of large intestine, rectum, prostate, breast, ovary, lung, and leukemia; weaker inverse associations were found for cancers of the pancreas, skin, and the bladder. In another study similar inverse correlations were observed between cancer mortalities at the above sites and the selenium concentrations in whole blood which had been collected from healthy human donors in the U.S. as well as several other countries.[130] Schrauzer et al. postulated that the cancer mortalities in the U.S. as well as other Western industrialized nations would markedly be reduced if the dietary selenium intakes were increased to about twice the current average amount supplied by the U.S. diet.[130]

In general the blood of cancer patients contains less selenium than the blood of healthy individuals.[131,132] It is not possible, however, to determine whether or not low blood selenium in cancer patients contributes to the disease or is an early effect of the anorexia or cachexia observed in cancer patients.

D. Mechanism of Action

Selenium may act through a selenoprotein, glutathione peroxidase, which uses hydrogen peroxide as a substrate. In this way fewer free radicals would be formed because they result of the interaction of hydrogen peroxide and cellular components. There is also a possibility that malonaldehyde may be increased through the peroxidation of unsaturated fat when selenium and vitamin E levels are low. Malonaldehyde has been shown to have tumor-initiating properties.[133]

Selenium may be decreasing metabolic pathways which produce precursors of DNA. Metabolic incorporation of (^{14}C) from (^{14}C) DMH into adenine and guanine (presumably via C, pathways) was reduced 69 to 72% in colon DNA of selenium-treated rats.[134] In addition, (^{3}H) thymidine incorporation was reduced 61 to 65%.

Ip has found that selenium does not affect the levels of circulating prolactin and estrogens,[135] hormones that are important for the development of DMBA-induced mammary tumors.

Russell et al. have found that selenium compounds are able to induce DNA repair synthesis as a measure of DNA damage in isolated rat liver cells. An enhancement of repair of BOP induced damage of pancreas DNA has been observed when the hamsters were pretreated with selenium.[136]

Selenium has been shown to enhance the effect of humoral immunity,[138] cell-mediated immunity,[139] as well as nonspecific immune effects.[140]

VIII. OTHER SUBSTANCES

Several other substances including indoles, aromatic isothiocyanates, methylated flavones, coumarins, and protease inhibitors are also able to inhibit carcinogenesis. It is possible that some of these substances may emerge as important chemopreventative substances.

REFERENCES

1. **Davies, R. E.**, Effect of vitamin A on 7,12-dimethylbenzanthracene-induced papillomas in rhino mouse skin, *Cancer Res.*, 27, 237, 1967.
2. **Bollag, W.**, Therapeutic effects of an aromatic retinoic acid analog on chemically induced benign and malignant epithelial tumors by vitamin A acid (retinoic acid), *Eur. J. Cancer*, 10, 731, 1972.
3. **Saffiotti, U., Montesano, R., and Sellakumar, A. R.**, Experimental cancer of the lung. Inhibition by vitamin A on the induction of tracheobronchial squamous metaplasia and squamous cell tumors, *Cancer Res.*, 20, 857, 1967.
4. **Cohen, S. M., Wittenberg, J. F., and Bryan, G. T.**, Effect of avitaminosis A and hypervitaminosis A on urinary bladder carcinogenicity of N-[4-(5-nitro-2-furyl)-2-thiazolyl] formamide, *Cancer Res.*, 36, 2334, 1976.
5. **Narisawa, T., Reddy, B. S., Wong, C. Q., and Weisburger, J. H.**, Effect of vitamin A deficiency on rat colon carcinogenesis by N-methyl-N-nitro-nitrosoguanidine, *Cancer Res.*, 36, 1379, 1976.
6. **Chu, E. W. and Malmgren, R. A.**, An inhibitory effect of vitamin A on the induction of tumors of forestomach and cervix in the Syrian hamster by carcinogenic polycyclic hydrocarbons, *Cancer Res.*, 25, 884, 1965.
7. **Merriman, R. L. and Bertram, J. S.**, Reversible inhibition by retinoids of 3-methylcholanthrene-induced neoplastic transformation on C3H/10T 1/2 clone 8 cells, *Cancer Res.*, 39, 1661, 1979.
8. **Bollag, W.**, Therapy of epithelial tumors with an aromatic retinoic acid analog, *Chemotherpy (Basel)*, 21, 236, 1975.
9. **Port, C. D., Sporn, M. B., and Kaufman, D. G.**, Prevention of lung cancer in hamsters by 13-cis-retinoic acid, *Proc. Am. Assoc. Cancer Res.*, 16, 21, 1975.
10. **Grubbs, C. J., Moon, R. C., and Sporn, M. B.**, Suppression of DMBA-induced mammary tumorigenesis by retinyl methyl ether, *Proc. Am. Assoc. Cancer Res.*, 17, 68, 1976.

11. **Sporn, M. B., Squire, R. A., Brown, C. C., Smith, J. M., Wenk, M. L., and Springer, S.,** 13-cis-Retinoic acid: inhibition of bladder carcinogenesis in the rat, *Science,* 195, 487, 1977.

12. **Busk, L. and Ahlborg, U. G.,** Retinol (vitamin A) as an inhibitor of the mutagenicity of aflatoxin B-1, *Toxicol. Lett.,* 6, 243, 1980.

13. **Huang, C. C., Hsueh, J. L., Chen, H. H., and Batt, T. R.,** Retinol (vitamin A) inhibits sister chromatid exchanges and cell cycle delay induced by cyclophosphamide and aflatoxin B-1 in Chinese hamster V 79 cells, *Carcinogenesis,* 3, 1, 1982.

14. **Hill, D. L. and Grubbs, C. J.,** Retinoids as chemopreventive and anticancer agents in intact animals (review), *Anticancer Res.,* 2, 111, 1982.

15. **Carruthers, C.,** The effect of carcinogens on the hepatic vitamin A stores of mice and rats, *Cancer Res.,* 2, 168, 1942.

16. **Basu, T. K., Donaldson, D., Jenner, M., Williams, D. C., and Sakula, A.,** Plasma vitamin A in patients with bronchial carcinoma, *Br. J. Cancer,* 33, 119, 1976.

17. **Graham, S., Mettlin, C., Marshall, J., Priore, R., Rzepka, T., and Shedd, D.,** Dietary factors in the epidemiology of cancer of the larynx, *Am. J. Epidemiol.,* 113, 675, 1981.

18. **Mettlin, C., Graham, S., and Swanson, M.,** Vitamin A and lung cancer, *J. Natl. Cancer Inst.,* 62, 1435, 1979.

19. **Sani, B. P. and Hill, D. L.,** Retinoic acid: a binding protein in chick embryo metatarsal skin, *Biochem. Biophys. Res. Commun.,* 61, 1276, 1974.

20. **Mettlin, C., Graham, S., Priore, R., Marshall, J., and Swanson, M.,** Diet and cancer of the esophagus, *Nutr. Cancer,* 2, 143, 1981.

21. **Hormozdiari, H., Day, N. E., Aramesh, B., and Mahboubi, E.,** Dietary factors and esophageal cancer in the Caspian littoral of Iran, *Cancer Res.,* 35, 3493, 1975.

22. **Graham, S., Schotz, W., and Martino, P.,** Alimentary factors in the epidemiology of gastric cancer, *Cancer,* 30, 927, 1972.

23. **Haenszel, W., Kurihara, M., Segi, M., and Lee, R. K. C.,** Stomach cancer among Japanese in Hawaii, *J. Natl. Cancer Inst.,* 49, 969, 1972.

24. **Bjelke, E.,** Dietary factors and epidemiology of cancer of the stomach and large bowel, *Aktuel Ernaehrungsmed. Klin. Prax. Suppl.,* 2, 10, 1978.

25. **Schuman, L. M., Mandell, J. S., Redke, A., Seal, U., and Halberg, F.,** Some selected features of the epidemiology of prostatic cancer: Minneapolis-St. Paul, Minnesota case-control study, 1976—1979, in *Trends in Cancer Incidence: Causes and Practical Implications,* Magnus, K., Ed., Hemisphere Publishing, New York, 1982, 345.

26. **Kark, J. D., Smith, A. H., and Hames, C. G.,** The relationship of serum cholesterol to the incidence of cancer in Evans County, Georgia, *J. Chronic Dis.,* 33, 311, 1980.

27. **Kark, J. D., Smith, A. H., and Hames, C. G.,** Serum retinol and the inverse relationship between serum cholesterol and cancer, *Br. Med. J.,* 284, 152, 1982.

28. **Hill, D. L. and Shih, T. W.,** Vitamin A compounds and analogues as inhibitors of mixed-function oxidases that metabolize carcinogenic polycyclic hydrocarbons and other compounds, *Cancer Res.,* 34, 564, 1974.

29. **Hirayama, T.,** The epidemiology of cancer of the stomach in Japan with special reference to the role of the diet, in *Proc. 9th Int. Congr.,* UICC Monograph Series Vol. 10, Harris, R. J. C., Ed., Springer-Verlag, New York, 1967, 37.

30. **Genta, V. M., Kaufman, D. G., Harris, C. C., Smith, J. M., Sporn, M. B., and Saffiotti, U.,** Vitamin A deficiency enhances the binding of benzopyrene to tracheal epithelial DNA, *Nature (London),* 247, 48, 1974.

31. **Verma, A. K., Shapas, B. G., Rice, H. M., and Boutwell, R. K.,** Correlation of the inhibition by retinoids of tumor-promotor-induced mouse ornithine decarboxylase activity of skin tumor promotion, *Cancer Res.,* 39, 419, 1979.

32. **Lowe, N. J. and Breeding, J.,** Retinoic acid modulation of ultraviolet light-induced epidermal ornithine decarboxylase activity, *J. Invest. Dermatol.,* 78, 121, 1982.

33. **Lotan, R. and Nicholson, G. L.,** Inhibitory effects of retinoic acid or retinyl acetate on the growth of untransformed, transformed, and tumor cells in vitro, *J. Natl. Cancer Inst.,* 59, 1712, 1977.

34. **Shamberger, R. J.,** Increase of peroxidation in carcinogenesis, *J. Natl. Cancer Inst.,* 48, 1491, 1972.

35. **Slaga, T. J. and Bracken, W. M.,** The effects of anti-oxidants on skin tumor initiation and aryl hydrocarbon hydroxylase, *Cancer Res.,* 37, 1631, 1977.

36. **Sadek, I. A. and Abdelmegid, N.,** Ascorbic acid and its effect on the skin of *Bufo regularis, Oncology,* 39, 399, 1982.

37. **Dunham, W. B., Zuckerkandl, E., Reynolds, R., Willoughby, R., Marcuson, R., Barth, R., and Pauling, L.,** Effects of intake of L-ascorbic acid on the incidence of dermal neoplasma induced in mice by ultraviolet light, *Proc. Natl. Acad. Sci. U.S.A.,* 79, 7532, 1982.

38. **Reddy, B. S., Hirota, N., and Katayama, S.,** Effect of sodium ascorbate on 1,2-dimethylhydrazine- or methylnitrosourea-induced colon carcinogenesis in rats, *Carcinogenesis,* 3, 1097, 1982.

39. **Jacobs, M. E. and Griffin, A. C.**, Effects of selenium on chemical carcinogenesis, *Biol. Trace Elem. Res.*, 1, 1, 1979.
40. **Jones, F. E., Komorowski, R. A., and Condon, R. E.**, Chemoprevention of 1,2-dimethylhydrazine-induced large bowel neoplasma, *Surg. Forum*, 32, 435, 1981.
41. **Wirth, P. J., Dybing, E., von Bahr, C., and Thorgeirsson, S.**, Mechanism of N-hydroxylacetylarylamine mutagenicity in the Salmonella test system. Metabolic activation of N-hydroxyphenacetin by liver and kidney fractions from rat, mouse, hamster and man, *Mol. Pharmacol.*, 18, 117, 1980.
42. **Shamberger, R. J., Corlett, C. L., Beaman, K. D., and Kasten, B. L.**, Antioxidants reduce the mutagenic effect of malonaldehyde and beta-propiolactone, *Mutat. Res.*, 66, 349, 1979.
43. **Khudoley, V., Malaveille, C., and Bartsch, H.**, Mutagenicity studies in Salmonella typhimurium on some carcinogenic N-nitramines in vitro and in the host-mediated assay in rats, *Cancer Res.*, 41, 3205, 1981.
44. **Guttenplan, J. B.**, Mechanisms of inhibition by ascorbate of microbial mutagenesis induced by N-nitroso compounds, *Cancer Res.*, 38, 2018, 1979.
45. **Kakatnur, M. G., Murray, M. L., and Correa, P.**, Mutagenic properties of nitrosated spermidine, *Proc. Soc. Exp. Biol. Med.*, 158, 85, 1978.
46. **DeFlora, S. and Picciotto, A.**, Mutagenicity of cimetidine in nitrite-enriched human gastric juice, *Carcinogenesis*, 1, 925, 1980.
47. **Neale, S. and Solt, A. K.**, The effect of ascorbic acid on the amine-nitrite and nitrosamine mutagenicity in bacteria injected into mice, *Chem. Biol. Interact.*, 35, 199, 1981.
48. **Lin, J. Y., Wang, H. I., and Yeh, Y. C.**, The mutagenicity of soy bean sauce, *Food Cosmet. Toxicol.*, 17, 329, 1979.
49. **Lee, K. Y., Choi, H. W., and Park, S. C.**, Bacterial mutation test for the detection of potential carcinogenicity of nitrite-treated Korean raw fishes, *Korean J. Biochem.*, 11, 31, 1979.
50. **Munkres, K. D.**, Aging of Neurospora Crassa. III. Lethality and mutagenicity of ferrous ions, ascorbic acid, and malondialdehyde, *Mech. Aging Dev.*, 10, 249, 1979.
51. **Schlegel, J. U., Pipkin, G. E., Nishmura, R., and Schultz, G. N.**, The role of ascorbic acid in the prevention of bladder tumor formation, *Trans. Am. Assoc. Genitourinary Surg.*, 61, 85, 1969.
52. **Boyland, E. and Williams, D. C.**, Metabolism of tryptophan in patients suffering from cancer of the bladder, *Biochem. J.*, 64, 578, 1965.
53. **Price, J. M., Wear, J. B., Brown, R. R., Satter, E. J., and Olson, C.**, Studies on etiology of carcinoma of the urinary bladder, *J. Urol.*, 83, 376, 1963.
54. **Benassi, C. A., Perissinotto, B., and Allegri, G.**, The metabolism of tryptophan in patients with bladder cancer and other urological diseases, *Clin. Chem. Acta.* 8, 822, 1963.
55. **Pipkin, G. E., Nishimura, R., Banowsky, L., and Schlegel, J. U.**, Stabilization of urinary-3-hydroxyanthranilic acid by the oral administration of L-ascorbic acid, *Proc. Soc. Exp. Biol. Med.*, 126, 702, 1967.
56. **Bruce, W. R., Varghese, A. J., Furrer, R., and Land, P. C.**, A mutagen in human feces, in *Origins of Human Cancer*, Hiatt, H. H., Watson, J. D., and Winsten, J. A., Eds., Cold Spring Harbor Laboratory, New York, 1977, 1641.
57. **Dion, P. W., Bright-See, E. B., Smith, C. C., and Bruce, W. R.**, The effect of dietary ascorbic acid and alpha-tocopherol on fecal mutagenicity, *Mutat. Res.*, 102, 27, 1982.
58. **Rao, B. G., McDonald, I. A., and Hutchison, D. M.**, Nitrite-induced volatile mutagens from normal human feces, *Cancer.* 47, 889, 1981.
59. **Sram, R. J., Dobias, L., Pastorkova, A., Rossner, P., and Janka, L.**, Effect of ascorbic acid prophylaxis on the frequency of chromosome aberrations in the peripheral lymphocytes of coal-tar workers, *Mutat. Res.*, 120, 181, 1983.
60. **Shoyab, M.**, Inhibition of the binding of 7,12-dimethylbenzanthracene to DNA of murine epidermal cells in culture by vitamin A and Vitamin C, *Oncology*, 38, 187, 1981.
61. **Mirvish, S. S., Wallace, L., Eagen, M., and Shubik, P.**, Ascorbate-nitrite reaction: possible means of blocking the formation of carcinogenic compounds, *Science*, 177, 65, 1972.
62. **Raineri, R. and Weisburger, J. H.**, Reductions of gastric carcinogens with ascorbic acid, *Ann. N.Y. Acad. Sci.*, 258, 181, 1975.
63. **Archer, M. C., Tannenbaum, S. R., Tan, T., and Weisman, M.**, Reaction of gastric carcinogens with ascorbic acid and its relation to nitrosamine formation, *J. Natl. Cancer Inst.*, 54, 1203, 1975.
64. **Chan, W. C. and Fong, Y. Y.**, Ascorbic acid prevents liver tumor production of aminopyrine and nitrite in the rat, *Int. J. Cancer*, 20, 268, 1970.
65. **Werner, B., Denzer, U., Mitschke, H., and Brassow, F.**, Vitamin C and Dunndarmkrebs. Eine Experimentelle studie, *Langenbecks Arch. Chir.*, 354, 101, 1981.
66. **Newberne, P. M.**, Carcinogenicity of aflatoxin-contaminated peanut meals, in *Mycotoxins in Foodstuffs*, Wogen, G. N., Ed., MIT Press, Cambridge, Mass., 1965, 187.
67. **Lombardi, B. and Shinozuka, H.**, Enhancement of 2-acetylaminofluorene liver carcinogenesis in rats fed a choline-devoid diet, *Int. J. Cancer*, 23, 565, 1979.

68. **Shinozuka, H., Katyal, S. L., and Lombardi, B.**, Azaserine carcinogenesis: organ susceptibility change in rats fed a diet devoid of choline, *Int. J. Cancer*, 22, 36, 1978.

69. **Takahashi, S., Lombardi, B., and Shinozuka, H.**, Progression of carcinogen-induced foci of gamma-glutamyltranspeptidase-positive hepatocytes to hepatomas in rats fed a choline-deficient diet, *Int. J. Cancer*, 29, 445, 1982.

70. **Shinozuka, H., Katyal, S. L., and Abanobi, S. E.**, Modifications of the promoting efficacy of a choline deficient diet by varying the type of dietary fat, *Proc. Am. Assoc. Cancer Res.*, 24, 409, 1983.

71. **Ghoshal, A. K. and Farber, E.**, Induction of liver cancer by a diet deficient in choline and methionine, *Proc. Am. Assoc. Cancer Res.*, 24, 388, 1983.

72. **Poirier, L. A., Mikol, Y. B., Hoover, K., and Creasia, D.**, Liver tumor formation in rats fed methyl-deficient, amino acid-defined diets with and without diethylnitrosamine initiation, *Proc. Am. Assoc. Cancer Res.*, 24, 384, 1983.

73. **Reddy, T. V., Ramanathan, R., Shinozuka, H., and Lombardi, B.**, Effects of dietary choline deficiency on the mutagenic activation of chemical carcinogens by rat liver fractions, *Cancer Lett.*, 18, 41, 1983.

74. **Rogers, A. E.**, Variable effects of a lipotrope deficient, high fat diet on chemical carcinogenesis in rats, *Cancer Res.*, 35, 2469, 1975.

75. **Giambarresi, L. I., Katyal, S. L., and Lombardi, B.**, Promotion of liver carcinogenesis in the rat by a choline-devoid diet: role of liver cell necrosis and regeneration, *Br. J. Cancer*, 46, 825, 1982.

76. **Shamberger, R. J.**, Relationships of selenium to Cancer. I. Inhibitory effect of selenium on carcinogenesis, *J. Natl. Cancer Inst.*, 44, 931, 1970.

77. **Jaffe, W.**, The influence of wheat germ oil on the production of tumors in rats by methylcholanthrene, *Exp. Med. Surg.*, 4, 278, 1949.

78. **Haddow, A. and Russell, H.**, The influence of wheat germ oil in the diet on the induction of tumors in mice, *Am. J. Cancer*, 29, 363, 1937.

79. **Epstein, S. S., Joshi, S., Andrea, J., Forsyth, J., and Mantal, N.**, The null effect of antioxidants on the carcinogenicity of 3,4,9,10-dibenzpyrene to mice, *Life Sci.*, 6, 225, 1967.

80. **Haber, S. L. and Wissler, R. W.**, Effect of vitamin E on carcinogenicity on methylcholanthrene, *Proc. Soc. Exp. Biol. Med.*, 111, 774, 1962.

81. **Harman, D.**, Dibenzanthracene induced cancer. Inhibiting effect of vitamin E, *Clin. Res.*, 17, 125, 1969.

82. **Ip, C.**, Dietary vitamin E intake and mammary carcinogenesis in rats, *Carcinogenesis*, 3, 1453, 1982.

83. **Narayan, K. A.**, Vitamin E deficiency and chemical carcinogenesis, *Experientia*, 26, 840, 1970.

84. **Shklar, G.**, Oral mucosal carcinogenesis in hamsters: inhibition by vitamin E, *J. Natl. Cancer Inst.*, 68, 791, 1982.

85. **Pauling, L., Willoughby, R., Reymonds, R., Blaisdell, B. E., and Lawson, S.**, Incidence of squamous cell carcinoma in hairless mice irradiated with ultraviolet light in relation to intake of ascorbic acid and of D. L. alpha-tocopheryl acetate, *Int. J. Vitam. Nutr. Res. Suppl.*, 79, 53, 1982.

86. **Prasad, K. N., Ramanujam, S., and Gaudreau, D.**, Vitamin E induces morphological differentiation and increases the effect of ionizing radiation on neuroblastoma cells in culture, *Proc. Soc. Exp. Biol. Med.*, 161, 570, 1979.

87. **Cook, M. G. and McNamara, P.**, Effect of dietary vitamin E on dimethylhydrazine-induced colonic tumors in mice, *Cancer Res.*, 40, 1329, 1980.

88. **Toth, B. and Patil, K.**, Enhancing effect of vitamin E on murine intestinal tumorigenesis by 1,2-dimethylhydrazine dihydrochloride, *J. Natl. Cancer Inst.*, 70, 1107, 1983.

89. **Shamberger, R. J., Baughman, F. F., Kalchert, S. S., and Willis, C. E.**, Carcinogen-induced chromosomal breakage decreased by antioxidants, *Proc. Natl. Acad. Sci.*, 70, 1461, 1973.

90. **Smalls, E. and Patterson, R. M.**, Reduction of benzopyrene induced chromosomal aberrations by DL-alpha-tocopherol, *Eur. J. Cell Biol.*, 28, 92, 1982.

91. **Dion, P. W., Bright-See, E. B., Smith, C. C., and Bruce, W. R.**, The effect of dietary ascorbic acid and alpha-tocopherol on fecal mutagenicity, *Mutat. Res.*, 102, 27, 1982.

92. **Kalina, L. M., Sardarly, G. M., and Alekperov, U. K.**, Antimutagenic effect of alpha-tocopherol on the frequency of gene mutations in Salmonella, *Genetika*, 15, 1880, 1979.

93. **Weisburger, J. H., Reddy, D. V. M., Hill, P., Cohen, L. A., Wynder, E. L., and Spingarn, N. E.**, Nutrition and cancer on the mechanisms bearing on causes of cancer of the colon, breast, prostate and stomach, *Bull. N.Y. Acad. Med.*, 56, 673, 1980.

94. **Abrams, A. A.**, Use of vitamin E in chronic cystic mastitis, *N. Engl. J. Med.*, 272, 1081, 1965.

95. **London, R. S., Solomon, D. M., London, E. D., Strummer, D., Bankowski, J., and Mair, P. P.**, Mammary dysplasia; clinical response and urinary excretion of 11-deoxyketosteroids and pregnanediol following alpha-tocopherol therapy, *Breast*, 4, 19, 1978.

96. **Ip, C. and Horvath, P.**, Synergistic effect of vitamin E and selenium in the chemoprevention of mammary carcinogenesis in rats, *Proc. Am. Assoc. Cancer Res.*, 24, 382, 1983.

97. **Konings, A. W. T. and Trieling, W. B.**, The inhibition of DNA synthesis in vitamin E-depleted lymphosarcoma cells by x-rays and cytostatics, *Int. J. Radiat. Biol.*, 31, 397, 1977.

98. **Gairola, C. and Chen, L. H.,** Effect of dietary vitamin E on the aryl hydrocarbon hydroxylase activity of various tissues in rat, *Int. J. Vitam. Nutr. Res.,* 52, 398, 1982.
99. **Malkinson, A. M.,** Review: putative mutagens and carcinogens in foods. III. Butylated hydroxytoluene (BHT), *Environ. Mutag.,* 5, 353, 1983.
100. **Batzinger, R. P., Ou, S. Y. L., and Bueding, E.,** Antimutagenic effects of 2(3)-tert-butyl-4-hydroxy-anisole and of antimicrobial agents, *Cancer Res.,* 38, 4478, 1978.
101. **Calle, L. M., Sullivan, P. D., Nettleman, M. O., Ocasio, I. J., Blazyk, J., and Jollick, J.,** Antioxidants and the mutagenicity of benzopyrene and some derivatives, *Biochem. Biophys. Res. Commun.,* 85, 351, 1978.
102. **McKee, R. H. and Tometsko, A. M.,** Inhibition of promutagen activation by the antioxidants butylated hydroxyanisole and butylated hydroxytoluene, *J. Natl. Cancer Inst.,* 63, 473, 1979.
103. **Shamberger, R. J., Corlet, C. L., Beamon, K. D., and Kasten, B. L.,** Antioxidants reduce the mutagenic effect of malonaldehyde and beta-propiolactone. IX. Antioxidants and cancer, *Mutat. Res.,* 66, 349, 1979.
104. **Shamberger, R. J., Baughmann, F. F., Kalchert, S. L., Willis, C. E., and Hoffman, G. C.,** Carcinogen-induced chromosomal breakage decreased by antioxidants, *Proc. Natl. Acad. Sci. U.S.A.,* 70, 1461, 1973.
105. Diet, Nutrition and Cancer, National Academy Press, Washington, D.C., 1982, 15-1.
106. **Wattenberg, L. W.,** Inhibitions of chemical carcinogens, in *Cancer: Achievements, Challenges and Prospects for the 1980's,* Vol. 1, Burchenol, J. H. and Oettgen, H. F., Eds., Grune & Stratton, New York, 1981, 517.
107. **Benson, S. M., Cha, Y. N., Bueding, E., Heine, H. S., and Talalay, P.,** Elevation of extrahepatic glutathione s-transferase and epoxide hydratase activities by 2(3)-*tert*-butyl-4-hydroxyanisole, *Cancer Res.,* 39, 2971, 1979.
108. **Cha, Y. N. and Bueding, E.,** Effect of 2(3)-*tert*-butyl-4-hydroxyanisole administration on the activities of several hepatic microsomal and cytoplasmic enzymes in mice, *Biochem. Pharmacol.,* 28, 1917, 1979.
109. **Cha, Y. N., Martz, F., and Bueding, E.,** Enhancement of liver microsome epoxide hydratase activity in rodents by treatment with 2(3)-*tert*-butyl-4-hydroxyanisole, *Cancer Res.,* 38, 4496, 1979.
110. **Lam, L. K. T., Fladmoe, A. V., Hochalter, J. B., and Wattenberg, L. W.,** Short time interval effects of butylated hydroxyanisole on the metabolism of benzopyrene, *Cancer Res.,* 40, 2824, 1980.
111. **Spier, J. L., Lam, L. K. T., and Wattenberg, L. W.,** Effects of administration to mice of butylated hydroxyanisole by oral intubation on benzopyrene-induced pulmonary adenoma formation and metabolism by benzopyrene, *J. Natl. Cancer Inst.,* 60, 605, 1978.
112. **Shamberger, R. J.,** Relationship of selenium to cancer. I. Inhibitory effect of selenium on carcinogenesis, *J. Natl. Cancer Inst.,* 44, 931, 1970.
113. **Clayton, C. C. and Baumann, C. A.,** Diet and azo dye tumors: effect of diet during a period when the dye is not fed, *Cancer Res.,* 9, 575, 1949.
114. **Griffin, A. C. and Jacobs, M. M.,** Effects of selenium on azo dye hepatocarcinogenesis, *Cancer Lett.,* 3, 177, 1977.
115. **Daoud, A. H. and Griffin, A. C.,** Effect of retinoic acid, butylated hydroxytoluene, selenium and sorbic acid on azo-dye hepatocarcinogenesis, *Cancer Lett.,* 9, 299, 1980.
116. **Balanski, R. M. and Hadsiolov, D. H.,** Influence of sodium selenite on the hepatocarcinogenic action of diethylnitrosamine in rats, *C. R. Acad. Bulg. Sci.,* 32, 697, 1979.
117. **Harr, J. R., Exon, J. H., Weswig, P. H., and Whanger, P. D.,** Relationship of dietary selenium concentration; chemical cancer induction; and tissue concentration of selenium in rats, *Clin. Toxicol.,* 8, 487, 1973.
118. **Jacobs, M. M., Jansson, B., and Griffin, A. C.,** Inhibitory effects of selenium on 1,2-dimethylhydrazine and methylazoxymethanol acetate induction of colon tumors, *Cancer Lett.,* 2, 133, 1977.
119. **Jacobs, M. M., Forst, C. F., and Beams, F. A.,** Biochemical and clinical effects of selenium on dimethylhydrazine-induced colon cancer in rats, *Cancer Res.,* 41, 4458, 1981.
120. **Schrauzer, G. N. and Ishmael, D.,** Effects of selenium and of arsenic on the genesis of spontaneous mammary tumors in inbred C_3H mice, *Ann. Clin. Lab. Sci.,* 4, 411, 1974.
121. **Thompson, H. J. and Taglaferro, A. R.,** Effect of selenium on 7,12-dimethylbenzanthracene-induced mammary tumorigenesis, *Fed. Proc.,* 39, 1117, 1980.
122. **Ip, C.,** Factors influencing the anticarcinogenic efficacy of selenium in dimethylbenzanthracene-induced mammary tumorigenesis in rats, *Cancer Res.,* 41, 2683, 1981.
123. **Birt, D. F., Lawson, T. A., Julius, A. D., Runice, C. E., and Saimasi, S.,** Inhibition by dietary selenium of colon cancer induced in the rat by bis(2-oxopropyl) nitrosamine, *Cancer Res.,* 42, 4455, 1982.
124. **Jacobs, M. M., Matney, T. S., and Griffin, A. C.,** Inhibitory effects of selenium on the mutagenicity of 2-acetylaminofluorene (AAF) and AAF Metabolites, *Cancer Lett.,* 2, 319, 1977.
125. **Adams, G., Martin, S., and Milner, J.,** Effects of selenium on the Salmonella/microsome mutagen test system, *Fed. Proc.,* 39, 790, 1980.

126. **Martin, S. E., Adams, G. H., Schillaci, M., and Milner, J. A.,** Antimutagenic effects of selenium on acridine orange and 7,12-dimethylbenzanthracene in the Ames Salmonella/microsomal system, *Mutat. Res.,* 82, 41, 1981.

127. **Ray, J. H., Altenburg, L. C., and Jacobs, M. M.,** Effects of sodium selenite and methyl methanesulphonate or N-hydroxy-2-acetylaminofluorene co-exposure on sister chromatid exchange production in human whole blood cultures, *Mutat. Res.,* 57, 359, 1978.

128. **Shamberger, R. J. and Willis, C. E.,** Selenium distribution and human cancer mortality, *CRC Crit. Rev. Clin. Lab. Sci.,* 2, 211, 1971.

129. **Shamberger, R. J., Tytko, S. A., and Willis, C. E.,** Antioxidants and cancer. VI. Selenium and age-adjusted human cancer mortality, *Arch. Environ. Health,* 31, 231, 1976.

130. **Schrauzer, G. N., White, D. A., and Schneider, C. J.,** Cancer mortality correlation studies. III. Statistical associations with dietary selenium intakes, *Bioinorg. Chem.,* 7, 23, 1977.

131. **Shamberger, R. J., Rukovena, E., Longfield, A. K., Tytko, S. A., Deodhar, S., and Willis, C. E.,** Antioxidants and cancer. I. Selenium in the blood of normals and cancer patients, *J. Natl. Cancer Inst.,* 50, 863, 1973.

132. **McConnell, K. P., Broghamer, W. L., Blotcky, A. J., and Hurt, O. J.,** Selenium levels in human blood and tissues in health and in disease, *J. Nutr.,* 105, 1026, 1975.

133. **Shamberger, R. J., Andreone, T. L., and Willis, C. E.,** Antioxidants and cancer. IV. Initiating activity of malonaldehyde as a carcinogen, *J. Natl. Cancer Inst.,* 53, 1771, 1974.

134. **Harbach, P. R. and Swenberg, J. A.,** Effects of selenium on 1,2-dimethyl-hydrazine metabolism and DNA alkylation, *Carcinogenesis,* 2, 575, 1981.

135. **Spallholz, J. E., Martin, J. L., Gerlach, M. L., and Heizerling, R. H.,** Injectable selenium: effect on the primary immune response of mice, *Proc. Soc. Exp. Biol. Med.,* 148, 37, 1975.

136. **Russell, G. R., Nader, C. J., and Patrick, E. J.,** Induction of DNA repair by some selenium compounds, *Cancer Lett.,* 10, 75, 1980.

137. **Lawson, T. and Birt, D.,** BOP induced damage of pancreas DNA and its repair in hamsters pretreated with selenium, *Proc. Am. Assoc. Cancer Res.,* 22, 93, 1981.

138. **Berenshtein, T. F.,** Effect of selenium and vitamin E on antibody formation in rabbits, *Zdrawookhr Boloruss,* 18, 34, 1972.

139. **Martin, J. L. and Spallholz, J. E.,** Selenium in the immune response, in *Proc. Symp. Selenium - Tellurium in the Environment,* Industrial Health Foundation, Pittsburgh, Pa., 1977, 204.

140. **Berenshtein, T. F.,** Stimulation of a nonspecific immunity in immunized rabbits by sodium selenite, *Ser. Biyal Navek,* 1, 87, 1973.

Chapter 4

NATURAL INHIBITORS OF CARCINOGENESIS: FERMENTED MILK PRODUCTS

D. R. Rao, S. R. Pulusani, and C. B. Chawan

TABLE OF CONTENTS

I. INTRODUCTION

It is well documented that nutrition is a major environmental factor in modulating carcinogenesis.[1] In the other chapters of this book, this relationship is elegantly discussed. It has been reported that processing of food such as charcoal broiling and smoking may lead to the production of mutagenic and carcinogenic compounds.[2] On the other hand, there are indications that food processing such as fermentation may result in the production of inhibitors of carcinogenesis. Recent studies show that fermented milk products indeed may be endowed with such properties.

Consumption of fermented foods is extensive in the world especially in the Orient.[3] Among the fermented foods, fermented milks have attracted the attention of both the laymen and the scientists for a long time as a healthful food. In 1908, Metchnikoff[4] postulated that the harmful effects of toxic amines (ptomaines) produced by intestinal putrefactive microorganisms could be minimized by establishing lactobacillus flora in the gut. He believed that the unusually long life span of the Balkan peasants is due to the consumption of large amounts of milk fermented by *Lactobacillus bulgaricus*. This hypothesis remained as a folklore and has been a controversial topic among the scientific and medical communities. Recent investigations, however, reveal that some strains of lactic acid bacteria used in the dairy industry for fermenting milk may possess some desirable properties. Recent investigations in our laboratory and elsewhere indicate that cultured dairy products and cultures used in fermenting milk may possess antitumorigenic activity.

II. EPIDEMIOLOGY

There is no direct epidemiologic evidence supporting the experimentally observed anticarcinogenic activity of fermented milk. Some epidemiological studies, however, have revealed a negative correlation between the consumption of dairy products and the incidence of certain types of cancer. In one epidemiologic study,[5] the intake of fiber, meat, and milk and levels of fecal bacteria in two Scandinavian populations with a four-fold difference in colon cancer incidence were analyzed. Correlation between dietary intake and the incidence of colon cancer suggested that the etiology of colon cancer may be multifactorial. Higher intakes of fiber and milk in the Finnish rural population with low incidence of colon cancer were suggested to have "a possible protective effect." Apparently the consumption of milk by Finnish rural population was four times higher than that of the Copenhagen men who showed a higher incidence of colon cancer. It was, however, not reported whether the data on milk consumption included any fermented dairy products such as yogurt and cheese. One relevant observation in this study is that the number of lactobacilli, some species of which are used in fermenting milk, was 1.6 log numbers per gram higher in the feces of Finnish population when compared to the lactobacilli counts in the feces of Copenhagen population. It is well documented that some of the lactobacilli used for fermenting milk can establish in animal and human GI tract.[6] The proportion of fermented milk consumed by these population samples is not available in the IARC[5] publication but the per capita consumption of fermented milk products in Finland is one of the highest in the world.[7,8] Calculation of per capita consumption of total fermented milk products (yogurt, buttermilk, sour milk, and Kefir) by Finnish population shows that they consume approximately 110 g of fermented milk products (excluding cheese) per person per day. Coincidentally, the incidence of colon cancer in Finland is one of the lowest among affluent countries.[9] Thus, the conclusion drawn by IARC[5] that milk and fiber might have a possible protective role against colon cancer needs further exploration with reference to the consumption of type of milk, especially fermented milk products. Other epidemiological studies also reported only "milk" consumption and no mention of consumption of fermented milk was made. Hirayama[10,11] in a series of studies conducted in Japan, found that the risk of stomach cancer was lowest in people consuming two glasses of milk per day. Supporting experimental data show that milk may play an important role in the prevention of gastric cancer caused by alkylating agents.[12] In contrast, a couple of other epidemiological interpretations incriminate a positive correlation between cancers of the colon, prostate,[13] and breast,[14] and milk consumption. Thus available information appears conflicting and often incomplete to associate the consumption of fermented milk with the incidence of any cancer. No retrospective case control studies or cohort studies dealing with the tumor inhibition activity of milk or fermented milk products are available. Clearly, undertaking a major epidemiological study in this area would be helpful to corroborate the compelling evidence from experimental work indicating antineoplastic activity of fermented milk products.

III. BIOCHEMICAL EPIDEMIOLOGY

A. Activity of Enzymes

Some enzymes of microbial and/or intestinal origin have been attributed to play an important role in the conversion of certain procarcinogens to carcinogens in intestines (see Goldin and Gorbach).[15] Among these enzymes, β-glucosidase, β-glucuronidase, nitroreductase, azoreductase, and (steroid) alpha dehydroxylase have been shown experimentally to convert some procarcinogens to carcinogens.[16-20]

Goldin and Gorbach[21] found that feeding of viable cultures of *Lactobacillus acidophilus* (one of the lactic cultures used in the fermentation of milk) as supplements to rats significantly lowered the activity of fecal bacterial β-glucuronidase, nitroreductase, and azoreductase in rats consuming meat diets. Goldin et al.[22] extended these observations onto humans. The addition of viable *Lactobacillus acidophilus* supplements to the diets containing red meat decreased the activity of fecal bacterial β-glucuronidase and nitroreductase activities significantly (Figure 1). Three weeks after the lactic culture supplement was discontinued, the activity of nitroreductase returned to the base line while the activity of β-glucuronidase still remained lower than the base line activity. The authors concluded that the metabolic activity of fecal microflora was influenced by *L. acidophilus* supplements. Ayebo et al.[23] confirmed these observations using a different type of *L. acidophilus* supplement. In a switch over design, these authors used two groups of geriatric subjects. One group received supplements of unfermented milk containing viable cells of *L. acidophilus* and the control group received unfermented milk free of *L. acidophilus* cells. The consumption of *L. acidophilus* cells significantly decreased the acitivity of β-glucosidase and β-glucuronidase. These observations provide grounds for further testing cultured and culture containing dairy products for their ability in lowering the activity of certain enzymes incriminated in the conversion of some procarcinogens to carcinogens.

B. Intestinal Microecology

A temporary death blow was dealt to Metchnikoff's hypothesis[4] when it was observed that *Lactobacillus bulgaricus* used in the preparation of the Bulgarian fermented milk does not implant in the human intestinal tract.[24] However, *L. acidophilus*, another lactic culture used in fermenting milk has been shown to implant in GI tract of humans.[6,23,25,26] Ayebo et al.[23] studied the effect of dietary supplementation of *L. acidophilus* cells in milk on human fecal microflora. The consumption of *L. acidophilus* cells significantly reduced the counts of fecal coliforms and increased the level of lactobacilli (Figures 2A and 2B). The high counts of lactobacilli were observed even after dietary supplementation was discontinued suggesting that survival, multiplication, and possible implantation of *L. acidophilus* has taken place. The authors interpreted that *L. acidophilus* has a beneficial effect on the intestinal microecological profile which favored the suppression of putrefactive organisms presumably

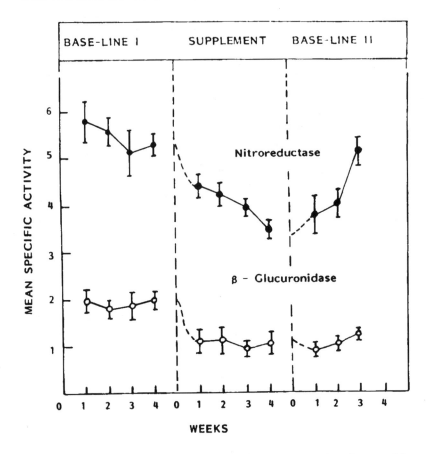

FIGURE 1.   Effect of dietary supplements of *L. acidophilus* on fecal nitroreductase and β-glucuronidase activities in humans consuming red meat. (From Goldin, B. R., Swenson, L., Dwyer, J., Sexton, M., and Gorbach, S. L., *J. Natl. Cancer Inst.*, 64, 255, 1980. With permission.)

involved in the production of putative carcinogens. In fact, some of the species of normal intestinal microflora have been shown to produce carcinogenic compounds.[15] In this regard, Mitsuoka[27] has demonstrated that *L. acidophilus* and *B. bifidum*, both used in dairy industry, when present in the intestines suppressed the formation of tumors produced by other intestinal microorganisms in germ-free mice. Similarly, *B. longum*, *L. acidophilus*, or *E. rectale* suppressed liver tumorigenesis in germ-free mice.[28] Rowland and Grasso[29] demonstrated that certain lactobacilli degrade the carcinogens dimethylnitrosamine and diphenylnitrosamine. Thus, it is apparent that interaction between lactic cultures and intestinal microbes may play an important role in tumorigenesis process. Some investigators were not able to demonstrate that *L. acidophilus* can be implanted in the human GI tract.[30,31] These contradictory results probably may be attributed to strain variation of *L. acidophilus* used. The existence of strain variation has been discussed by Klaenhammer.[32] He demonstrated that colonial variations, storage stability, and freeze damage can affect the implantation of *L. acidophilus* cells. Nonetheless, the implantation of lactic culture in the gut may not be a prerequisite for affecting the microecology of the gut. Other lactic cultures such as *L. bulgaricus* and *S. thermophilus* produce potent antimicrobial compounds.[33] Some antibiotics are known to possess antitumorigenic activity but it is not known whether antibiotics produced by lactic cultures possess any antitumorigenic activity. However, the presence of antibiotics in fermented milk might alter the GI microecology or the implanted/digested cultures might produce compounds that are antagonistic to the action of carcinogens.[34]

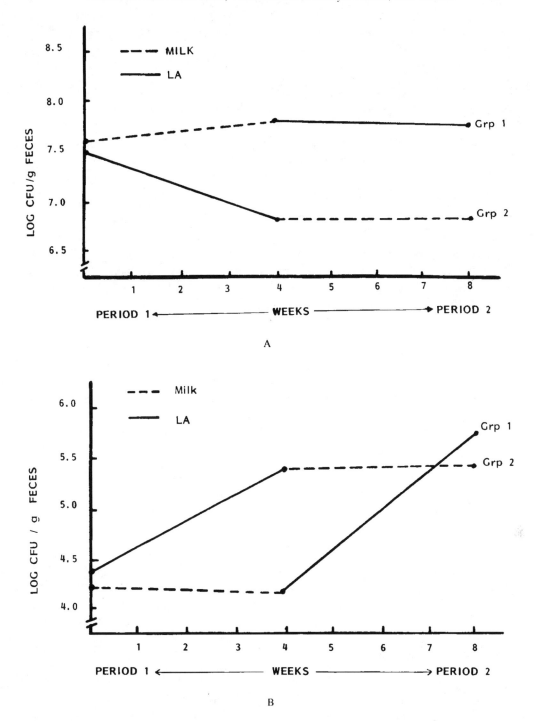

FIGURE 2. Mean fecal coliform count (A) and lactobacillus count (B) for two groups of subjects given milk and unfermented *L. acidophilus* milk (LA). (From Ayebo, A. D., Angelo, I. A., and Shahani, K. M., *Milchwissenschaft*, 35, 730, 1980. With permission.)

IV. ANIMAL MODEL AND IN VITRO STUDIES

Major evidence on the antineoplastic activity of fermented milk products comes from animal and in vitro tumor model studies. Tumor models used to test the antitumorigenic activity of cultured dairy products or of the lactic cultures used in preparing dairy products include colon tumors, sarcoma-180, Ehrlich ascitis tumor, AKATOL intestinal carcinoma, Lewis - LLC lung carcinoma, plasmacytoma MOPC-315, melanosarcoma B-16, Leukemia, KA-line cells, and Hi-line cells.

In 1962 Bogdanov et al.[35] reported that extracts of *L. bulgaricus*, one of the cultural organisms used in yogurt preparation, exhibited significant antitumor activity against sarcoma-180 and solid form of Ehrlich carcinoma. Five years later, Texim Enterprise Economique d'Etat[36] reported that extracts prepared from various lactic cultures used in dairy industry (*L. acidophilus* var. *tumoronecroticans*, *L. bulgaricus* var. *tumoronecroticans*, *L. casei* var. *tumoronecroticans*, and *L. helveticus* var. *tumoronecroticans*) inhibited the growth of implanted sarcoma-180 and Ehrlich carcinoma. Bogdanov and associates[37,38] were able to isolate three glycopeptides from the cell walls of *L. bulgaricus* which exhibited substantial antitumorigenic activity against sarcoma-180 and solid Ehrlich ascites tumor. They found that i.v. or i.p. injections of these three glycopeptides inhibited the growth of several syngenic and allogenic tumors in mice and prolonged the survival time of leukemic animals. Of the tumor models studied, the glycopeptides showed maximum inhibitory activity against sarcoma-180. Since the glycopeptides were ineffective against in vitro tumor models and some of the cured animals exhibited subsequent immunity, the authors interpreted that the glycopeptide fragments (blastolysin) from the cell wall of *L. bulgaricus* stimulated the host's immune system. Further evidence on the stimulation of host's immune system by lactic organisms has been provided by Kato et al.[39] They studied the effect of *L. casei* YIT 9018, an organism used in the preparation of yakult (yogurt) in Japan, on the growth of transplantable allogenic and syngenic mouse tumors. Intraperitoneal injection of *L. casei* cells resulted in a significant increase in the survival time of ICR mice implanted with sarcoma-180 and BDFI mice implanted with L1210 leukemia cells. When injected i.v., *L. casei* markedly inhibited the growth of subcutaneously inoculated sarcoma-180 and methylcholanthrene-induced syngenic MCA K-1 tumor in BALB/c mice. Furthermore, oral administration of *L. casei* also inhibited the growth of sarcoma-180 inoculated s.c. in ICR mice. The antitumor activity of *L. casei* was reduced in mice treated with carrageenan, an inhibitor of macrophage function (Figure 3). However, the antitumor activity was not affected in T-cell deficient athymic nude mice. The authors concluded that the antitumor activity of *L. casei* used in the study was macrophage dependent. Mitsuoka[27] suggested that lactobacilli inhibit carcinogenesis by (1) inactivating or inhibiting the formation of carcinogenic compound and (2) by stimulating host's immune mechanism. Thus, it appears that lactic bacteria used in fermenting milk may suppress the growth of certain tumors by enhancing the host's immune response.

Other work with lactobacilli indicates that two strains of *L. bulgaricus* and one strain of *L. acidophilus* (DDS-1) possessed marked antineoplastic activity against sarcoma-180.[40] Partially purified fractions from *L. acidophilus* (DDS-1) produced morphological changes and markedly inhibited the growth of KB-line cells in vitro. Only one of the fractions was active against sarcoma-180 and none were active against sarcoma-180 and Ridgeway osteogenic sarcoma.[40]

Direct antitumor activity of the fermented milk products has been demonstrated by a series of experiments conducted by Shahani and associates. In 1973 Reddy et al.[41] reported that when male swiss mice with Ehrlich ascites tumor transplants were fed *ad libitum* yogurt mixed with drinking water for 8 days, tumor cell proliferation was reduced by 28%. They confirmed these observations subsequently.[42] Shahani and associates[43,44] demonstrated that lactic acid (a major metabolite produced by lactic bacteria in yogurt and other fermented milk products), fresh unfermented milk, and unfermented colostrum had no effect on proliferation of Ehrlich ascites tumor cells in mice, while all of the lactic fermented milks and colostrums displayed significant inhibition against Ehrlich ascites tumor cell proliferation

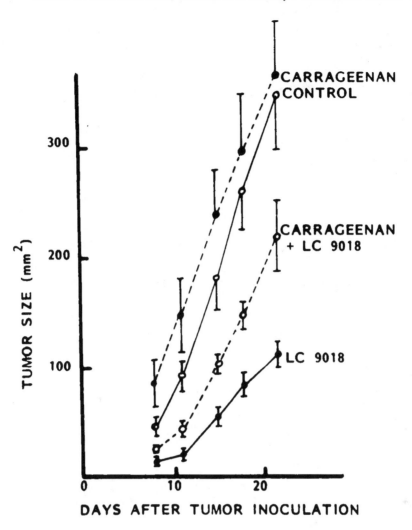

FIGURE 3. Effect of carrageenan on the antitumor activity of LC 9018. Sarcoma-180 was inoculated s.c. into ICR mice on day 0. Carrageenan (20 mg/kg) was injected i.v. on days +1, +2, +3, +4, and +5. (From Kato, I., Kabayashi, S., Yokokura, TY., and Mutai, M., *Gann*, 72, 517, 1981. With permission.)

when fed to mice. Lactic cultures used in these studies for fermenting milk and colostrum included *L. acidophilus*, *L. bulgaricus*, and a combination of *L. bulgaricus* and *Streptococcus thermophilus*. Mitsuoka[27] also demonstrated that feeding of fresh or pasteurized yogurt inhibited the Ehrlich ascites tumor cell growth by 30%. Further, work at the University of Nebraska demonstrated that the antitumorigenic compounds were loctated in the cell wall fraction (Table 1) confirming the observations of Bogdanov et al.[37,38] Parodoxically, in another study,[45] they observed that the antitumor compounds produced by lactic cultures in yogurt were dialyzable and of small molecular weight (14,000 daltons). The yogurt dialyzate inhibited the tumor growth by 39% when injected i.v. into mice implanted with Ehrlich ascites tumor cells. The inhibition was also significant when the yogurt dialyzate was fed orally. This paradox was explained on the assumption that there is a continuous autolysis of aging *L. bulgaricus* and *S. thermophilus* (cultures used in yogurt preparation) cells resulting in disintegration of cell wall components and their concomittant release into the medium.[44] Scioli[46] examined the anticarcinogenic activity of yogurt in a tumor model system involving the repair of carcinogen-induced DNA damage in hamsters (Figure 4). It was concluded that

Table 1
EFFECT OF I.P. IMPLANTATION OF CULTURE BROTH, YOGURT CULTURE, SOLUBLE- AND INSOLUBLE-SONICATED CULTURE CELLS, AND INDIVIDUAL CULTURES OF *L. BULGARICUS* AND *S. THERMOPHILUS*

| | Tumor cells (10^6/mouse) | | Inhibition |
Material implanted	Control	Test	(%)
Culture broth	13.9	14.4	0
Whole yogurt cells	14.3	6.0	58.0
Sonicate/soluble	15.4	15.3	0
Sonicate/insoluble	15.4	9.8	36.2
L. bulgaricus cells	11.6	7.7	33.6
S. thermophilus cells	11.6	8.5	26.7

From Friend, B. A., Farmer, R. E., and Shahani, K. M., *Milchwissenschaft*, 37, 708, 1982. With permission.

yogurt- and acidophilus milk-fed hamsters were able to repair DNA damage more quickly than animals receiving the control diets. The author, however, did not discuss the possiblity of yogurt components protecting DNA from damage by the injected carcinogen. This protective mechanism appears more plausible since certain compounds such as antioxidants scavenge the carcinogens before they reach the target site.[47] In fact, our recent studies indicate that *S. thermophilus* produces (or enhances the activity of) antioxidant-like compounds when cultured in milk (Figure 5).[34] Thus, it appears that in addition to enhancing the immune response of the host, yogurt cultures may protect the DNA damage by certain carcinogens.

There have been very few animal model studies in which the effect of feeding fermented milk on the incidence of tumors of clinical significance has been investigated. Goldin and Gorbach[48] studied the effect of *L. acidophilus* dietary supplements on dimethyl hydrazine-induced intestinal cancers in rats. The colon tumor incidence after a 20-week induction period was lower in rats receiving *L. acidophilus* supplements (40 vs. 77%) than in control rats. Recently, we found that feeding supplements of milk fermented by *L. bulgaricus* and *S. thermophilus* to rats significantly altered the incidence of chemically induced tumors.[49] In this study, weanling Fisher-344 rats were fed chow plus skim milk (SM), chow plus SM fermented by *S. thermophilus*, chow plus SM fermented by *L. bulgaricus*, or chow plus water until sacrifice at 36 weeks, or before if moribund. Colon tumors were induced by s.c. injections of 1,2-dimethylhydrazine hydrochloride during weeks 3 through 22. The control (chow + water) group received saline injections. The survival rate of the rats fed fermented milks was significantly higher than that of the rats fed unfermented milk (Figure 6). The latter had a significantly higher incidence of ear-duct tumors than the rats receiving fermented milk. The percentage of rats showing colon tumors was similar among all three experimental groups. However, the rats on *S. thermophilus* milk had the lowest percentage of malignant colon tumors among the three experimental groups (Table 2). The rats receiving fermented milk had a significantly higher incidence of small-intestine tumors than those receiving unfermented milk. Results indicated that the feeding of fermented milks altered the metabolism of 1,2-dimethylhydrazine and shifted the target organ from the ear duct to the small intestine. In addition, the colon tumor distribution for the fermented milk groups appeared to shift toward the anus. In another study,[50] supplements of fermented milk (fermented by *S. thermophilus*, *L. bulgaricus*, or *L. acidophilus*) were fed to 200 Fisher-344 rats either

ELUTION RATE CONSTANTS VS. TIME

(Diets prefed 20 days)

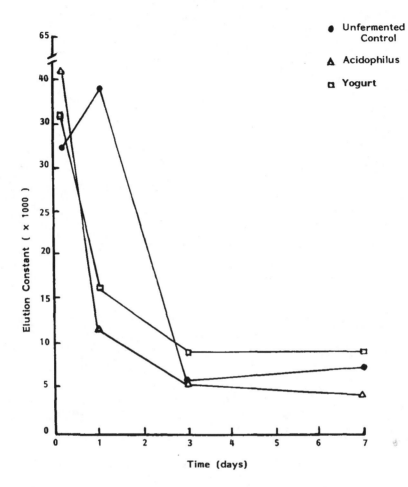

FIGURE 4.   Elution rate constants for DNA from hamsters fed an unfermented control diet, acidophilus diet and yogurt diet and injected with *N*-nitrosobis (2-oxypropyl) amine (BOP). (From Scioli, S. E., Master's thesis, University of Nebraska, 1982. With permission.)

during tumor initiation (induced by azoxymethane) or tumor promotion stages. After 50 weeks, the survival rate for rats fed fermented milks was numerically but not significantly higher than that of the rats fed unfermented milk. Rats fed unfermented milk showed the highest incidence of intestinal tumors (initiation stage 35%, promotion stage 33%) while rats fed milk fermented by *L. bulgaricus* showed the lowest incidence (initiation stage 26%; promotion stage 21%). Similar results were observed for colon tumors. A nonsignificant reduction in this study might indicate that antineoplastic action of lactic culture may be additive and is exerted at both initiation and promotion stages.

V. SUMMARY

Although there is evidence in the literature suggesting antitumorigenic action of cultured (culture containing) dairy products and lactic cultures used in preparation of these products, the evidence is not unequivocal. There is no direct epidemiologic evidence linking the consumption of fermented milk products and reduced incidence of certain tumors. Some

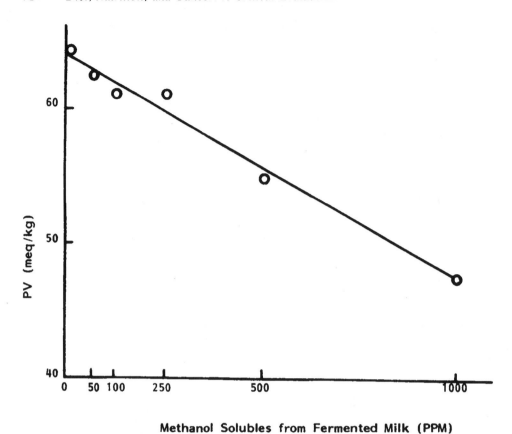

FIGURE 5.   Antioxidant activity of methanol solubles from milk fermented by *Streptococcus thermophilus*. Methanol solubles were added to 5 g aliquotes of corn oil stored at room temperature under fluorescent lights. The peroxide value (PV) of the corn oil was measured after 20 days of storage.

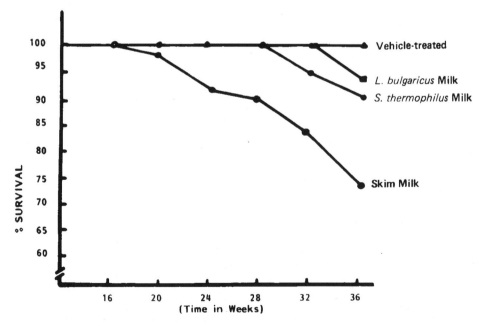

FIGURE 6.   Survival rate of rats injected with DMH and fed supplements of different fermented milks. (From Shackelford, L. A., Rao, D. R., Chawan, C. B., and Pulusani, S. R., *Nutr. Cancer*, 5, 159, 1983. With permission.)

Table 2
DISTRIBUTION OF TUMORS AMONG RATS TREATED WITH DMH AND FED FERMENTED AND NONFERMENTED MILK (N = 28 FOR EACH GROUP)

Tumor site	Skim milk group No. of rats positive	No. of tumors	S. thermophilus group No. of rats positive	No. of tumors	L. bulgaricus group No. of rats positive	No. of tumors
Earduct	15[a]	15[c]	7[b]	7[d]	8[b]	8[c]
Small intestine	9[a]	13	15[b]	16	10[a]	15
Colon	24	58	24	65	26	66
Benign		8		9		8
Malignant		10		6		8

[a,b] Numbers in the same row bearing unlike superscripts are significantly different ($p < 0.01$); comparisons are for positive rats vs. positive rats and no. of tumors vs. no. of tumors for three treatments.

[c,d] Numbers in the same row bearing unlike superscripts are significantly different ($p < 0.05$).

From Shackleford, L. A., Rao, D. R., Chawan, C. B., and Pulusani, S. R., *Nutr. Cancer*, 5, 159, 1983. With permission.

indirect epidemiologic evidence, however, indicates that milk, some of which is consumed in fermented form in areas with low incidence of certain cancers, may have a protective role. Biochemical epidemiologic evidence strongly suggests that fermented milk products may possess antitumorigenic properties. Most of the compelling evidence is from animal and in vitro model studies. Several tumor models have been shown to be inhibited by cultured dairy products or cultures used in fermenting milk. Available literature indicates that cultured dairy products (or cultures used therein) inhibit tumorigenesis by (1) enhancing the host's immune response, (2) suppressing the growth of intestinal bacteria incriminated in producing putative carcinogen, and/or (3) producing antitumorigenic compounds which might scavenge the carcinogens. Thus, the available evidence provides ample justification for intense future research in characterizing the anticarcinogenic activity of fermented milk products.

ACKNOWLEDGMENTS

Some of the work reported herein has been supported by grants from USDA (SEA/CR 95-113). The helpful discussion with Dr. B. S. Reddy in this work is appreciated.

REFERENCES

1. National Academy of Sciences, Diet, Nutrition and Cancer, Committee on Diet, Nutrition and Cancer, Assembly of Life Sciences, National Research Council, Washington, D. C., National Academy Press, 1982.
2. **Weisburger, J. H. and Spingarn, N. E.**, Mutagens as a function of mode of cooking of meats, in *Naturally Occurring Carcinogens-Mutagens and Modulators of Carcinogenesis*, Miller, E. C., Miller, J. A., Hirona, I., Sugimura, T., and Takayama, S., Eds., University Park Press, Baltimore, Md., 1979, 177.
3. **Pederson, C. S.**, *Microbiology of Food Fermentations*, AVI Publishing, Westport, Conn., 1978.
4. **Metchnikoff, E.**, *The Prolongation of Life: Optimistic Studies*, G. P. Putman's Sons, New York, 1908.
5. IARC on cancer intestinal microecology group, Dietary fiber, transit-time, fecal bacteria, steroids, and colon cancer in two Scandinavian populations, *Lancet*, 2, 207, 1977.

6. **Shahani, K. M. and Ayebo, A. D.**, Role of dietary lactobacilli in gastro-intestinal microecology, *Am. J. Clin. Nutr.*, 33, 2448, 1980.

7. **Renner, E.**, *Milk and Dairy Products in Human Nutrition*, VV-GmbH Volkswirtschaftlicher, Verlag, Federal Republic of Germany, 1983.

8. **Tamime, A. Y. and Deeth, H. C.**, Yogurt: technology and biochemistry, *J. Food Prot.*, 43, 939, 1980.

9. **Wynder, E. L.**, Epidemiology of large bowel cancer, *Cancer Res.*, 35, 3388, 1975.

10. **Hirayama, T.**, An epidemiological study on the effect of diet, especially of milk, on the incidence of stomach cancer, Abstract, 9th Int. Cancer Congr., Tokyo, 1966, P713.

11. **Hirayama, T.**, Diet and cancer, *Nutr. Cancer*, 1, 67, 1979.

12. **Jaquet, J., Huynk, C. H., and Saint, S.**, Nutrition and experimental cancer. Milk. Heba, C. R., Ed., *Seanc. Acad. Agric. Fr.*, 54, 112, 1977.

13. **Correa, P.**, Nutrition and cancer: epidemiologic correlation, in, *Nutrition and Cancer*, Newell, G. R. and Ellison, N. M., Eds., Raven Press, New York, 1981, 1.

14. **Gaskill, S. P., McGuire, W. L., Osborne, C. K., and Stern, M. P.**, Breast cancer mortality and diet in the United States, *Cancer Res.*, 39, 3628, 1979.

15. **Goldin, B. R. and Gorbach, S. L.**, Microbial factors and nutrition in carcinogenesis, in, *Advances in Nutritional Research*, Vol. 2, Draper, H. H. Ed., Plenum Press, New York, 1979, 129.

16. **Fisher, L. J., Millburn, P., and Smith, R. L.**, The fate of ^{14}C-stilbesterol in the rat, *Biochem. J.*, 100, 69, 1966.

17. **Hill, M. J., Draser, B. S., Aries, V., Crowithez, J. S., Hawksworth, G. M., and Williams, R. E. O.**, Bacteria and etiology of cancer of the large bowel, *Lancet*, 1, 95, 1971.

18. **Laqueur, G. L. and Spatz, M.**, Toxicology of cycasin, *Cancer Res.*, 28, 2262, 1968.

19. **Mastramarino, A., Reddy, B. S., and Wynder, E. L.**, Metabolic epidemiology of colon cancer: Enzyme activity of fecal flora, *Am. J. Clin. Nutr.*, 29, 1455, 1976.

20. **Reddy, B. S. and Wynder, E. L.**, Large bowel carcinogenesis — fecal constituents of populations with diverse incidence rates of colon cancer, *J. Natl. Cancer Inst.*, 50, 1437, 1973.

21. **Goldin, B. R. and Gorbach, S. L.**, Alterations in fecal microflora enzymes related to diet, age, lactobacillus supplements, and dimethylhydrazine, *Cancer*, 40, 2421, 1977.

22. **Goldin, B. R., Swenson, L., Dwyer, J., Sexton, M., and Gorbach, S. L.**, Effect of diet and *Lactobacillus acidophilus* supplements on human fecal bacterial enzymes, *J. Natl. Cancer Inst.*, 64, 255, 1980.

23. **Ayebo, A. D., Angelo, I. A., and Shahani, K. M.**, Effect of ingesting *Lactobacillus acidophilus* milk upon fecal flora and enzyme activity in humans, *Milchwissenschaft*, 35, 730, 1980.

24. **Rettger, L. F. and Cheplin, H. A.**, Therapeutic application of *Bacillus acidophilus*, *Proc. Soc. Exp. Biol. Med.*, 19, 72, 1921.

25. **Gilliland, S. E., Speck, M. L., Nauyok, G. F., Jr., and Giesbrecht, F. G.**, Influence of consuming nonfermented milk containing *Lactobacillus acidophilus* on fecal flora of healthy males, *J. Dairy Sci.*, 61, 1, 1978.

26. **Speck, M. L.**, Interactions among lactobacilli and man, *J. Dairy Sci.*, 59, 338, 1976.

27. **Mitsuoka, T.**, Intestinal flora and cancer, 2nd Annu. Natl. Symp. for Lactic Acid Bacteria and Health, Korea, 1981.

28. **Mizutani, T. and Mitsuoka, T.**, personal communication, 1983.

29. **Rowland, I. R. and Grasso, P.**, Degradation of N-nitrosamines by intestinal bacteria, *Appl. Microbiol.*, 29, 7, 1975.

30. **Paul, D. and Hoskins, L. C.**, Effect of oral lactobacillus feedings on fecal lactobacillus counts, *Am. J. Clin. Nutr.*, 25, 763, 1972.

31. **Robbins-Browne, R. M. and Levine, M. M.**, The fate of ingested lactobacilli in the proximal small intestine, *Am. J. Clin. Nutr.*, 34, 514, 1981.

32. **Klaenhammer, T. R.**, Microbiological considerations in selection and preparation of Lactobacillus strains for use as dietary adjuncts, *J. Dairy Sci.*, 65, 1339, 1982.

33. **Pulusani, S. R., Rao, D. R., and Sunki, G. R.**, Antimicrobial activity of lactic cultures: partial purification and characterization of antimicrobial compound(s) produced by *Streptococcus thermophilus*, *J. Food Sci.*, 2, 575, 1979.

34. **Rao, D. R., Pulusani, S. R., and Chawan, C. B.**, Antioxidant activity of milk and milk fermented by *Streptococcus thermophilus*, Food Science and Human Nutrition Section Abstracts, 81st SAAS Annual Meetings, Volume 21, 10, 1984.

35. **Bogdanov, I. G., Popkhristov, P., and Marinov, L.**, Anticancer effect of antibioticum bulgaricum on Sarcoma-180 and the solid form of Ehrlich carcinoma, Abstr. 8th Int. Cancer Congr., Moscow, 1962, 364.

36. Texim Enterprise Economique d-Etat, Anticarcinogens Microbial Extract, Neth, Patent Application, 6,607,345, 1967.

37. **Bogdanov, I. G., Dalev, P. G., and Gurevich, A. I.**, Antitumor *glycopeptides* from *Lactobacillus bulgaricus* cell walls, *FEBS Lett.*, 57, 259, 975.

38. **Bogdanov, I. G., Velichkov, V. T., and Gurevich, A. I.,** Antitumor action of glycopeptides from the cell wall of *Lactobacillus bulgaricus, Bull. Exp. Biol. Med.,* 84, 1750, 1978.

39. **Kato, I., Kabayashi, S., Yokokura, T., and Mutai, M.,** Antitumor activity of *Lactobacillus casei* in mice, *Gann,* 72, 517, 1981.

40. **Shahani, K. M.,** Isolation and Study of Anticarcinogenic Agents from Lactobacillus Fermented Food System, Final Report, Damon Runyan Memorial Fund for Cancer Research, Inc., 1969.

41. **Reddy, G. V., Shahani, K. M., and Banerjee, M. R.,** Inhibitory effect of yogurt on Ehrlich ascites tumor cell proliferation, *J. Natl. Cancer Inst.,* 50, 815, 1973.

42. **Reddy, G. V., Friend, B. A., Shahani, K. M., and Farmer, R. E.,** Antitumor activity of yogurt components, *J. Food Prot.,* 46, 8, 1983.

43. **Friend, B. A., Farmer, R. E., and Shahani, K. M.,** Effect of feeding and intraperitoneal implantation of yogurt culture cells on Ehrlich ascites tumor, *Milchwissenschaft,* 37, 708, 1982.

44. **Shahani, K. M. and Friend, B. A.,** The anti-neoplasm effects of lactobacilli, in Proc. 3rd Korean Public Health Assoc. Biennial Seminar, Korea, 1983, 2.

45. **Ayebo, A. D., Shahani, K. M., and Dam, R.,** Antitumor component(s) of yogurt: fractionation, *J. Dairy Sci.,* 64, 2318, 1981.

46. **Scioli, S. E.,** The Influence of Acidophilus Milk and Yogurt Diets on the Repair of Carcinogen-Induced DNA Damage, Master's thesis, University of Nebraska, 1982.

47. **Wattenberg, L. W.,** Inhibitors of chemical carcinogens, in *Cancer: Achievements, Challenges and Prospects for the 1980's,* Vol. 1, Burchenal, J. H. and Oettgen, H. F., Eds., Grune & Stratton, New York, 1981, 517.

48. **Goldin, B. R. and Gorbach, S. L.,** Effect of *Lactobacillus acidophilus* dietary supplements on 1,2-dimethylhydrazine hydrochloride-induced intestinal cancer in rats, *J. Natl. Cancer Inst.,* 64, 263, 1980.

49. **Shackelford, L. A., Rao, D. R., Chawan, C. B., and Pulusani, S. R.,** Effect of feeding fermented milk on the incidence of chemically-induced colon tumors in rats, *Nutr. Cancer,* 5, 159, 1983.

50. **Rao, D. R., Pulusani, S. R., and Omotayo, O.,** Effect of feeding fermented milk on the incidence of chemically-induced intestinal tumors in rats, paper presented at the Seminar on Fermented Milks, IDF, Avignon, France, May 14 to 16, 1984.

Chapter 5

EPIDEMIOLOGIC FINDINGS RELATING DIET TO CANCER OF THE ESOPHAGUS

Curtis Mettlin

TABLE OF CONTENTS

I. INTRODUCTION

Cancer of the esophagus is a malignancy that exhibits a pattern of occurrence that has attracted the attention of many researchers interested in the dietary and nutritional epidemiology of cancer. The esophagus is in such close contact with the dietary environment that it is intuitive that the search for environmental factors would focus on this category of risk. Furthermore, the geographic pathology of esophageal cancer reveals differences in risk by region so great that the cultures of the populations at high or low risk would be suspect in any attempt to understand the etiology of this cancer. Dietary customs, being such distinctive features of different cultures, might reasonably explain the geographical patterns of risk.

One would hope that an examination of the common risk factors for a disease, observed in different cultural contexts, would lead to a parsimonious description of the etiology of the disease. However, study of the dietary and other lifestyle factors that characterize different populations at high risk of cancer of the esophagus suggests several distinct, and plausible, etiologies for the disease. In this review, we examine the results of epidemiologic studies of dietary risk factors carried out in diverse regions at high risk of esophageal cancer. The emphasis of the review will be on studies of risk in human populations but, evidence from in vitro and other *in vivo* models may be relevant and will be included selectively. In the concluding section, the diverse and, sometimes conflicting, data will be reviewed to assess what common features may be evident and to suggest promising lines of inquiry.

II. HIGH-RISK POPULATIONS

Anecdotal reports of regional pockets of high esophageal cancer incidence have occurred sporadically for years but, it is only within the last 2 decades that rigorous and systematic investigations of populations at high risk have appeared in the literature. These reports come from multidisciplinary investigations conducted principally in seven geographically and culturally distinct populations including:

1. Asian Turkomans and the Turkic and Persian-speaking populations of northeastern Iran
2. The Chinese of the Taihang Mountain regions and Linxian Province of the People's Republic of China
3. Native populations of eastern and southern Africa
4. Males in the Brittany region of France
5. Black males in Washington D.C.
6. Singapore Chinese
7. The population of Puerto Rico

Although the most intensive study has occurred in these particular areas of exceptionally high incidence, other studies have been conducted among populations of more moderate risk and, some evidence also may be derived from these investigations.

III. CATEGORIES OF DIETARY RISK

Esophageal carcinogenesis has been attributed by different investigators to excesses and deficits of dietary exposure that may act through several different mechanisms. Some of these mechanisms are those hypothesized to affect the processes of cancer formation in other, related organs. For example, gastric cancer shares many epidemiologic features with esophageal cancer and, although the tissues are acknowledged to be distinct, it is possible that the two conditions share a dietary etiology. Thus, exposure to nitrates and nitroso compounds have been suspect. Because vitamins, such as A and C, have been linked to several different tumors of epithelial origin, these too have been studied in association with cancer of the esophagus. Several of the regions with high rates happen to be undeveloped and have primitive and often unsanitary food storage facilities. Mutagens or carcinogens from molds, from pickling or, from smoked preservation of foods are suspect. Finally, because of the relatively direct exposure of the esophagus to undigested dietary components, the role of direct physical trauma in the form of thermal or mechanical irritation has been hypothesized and investigated.

Each of these and some other categories of dietary risk will be examined herein. Importantly absent from this chapter is any direct examination of the roles of tobacco and alcohol in esophageal carcinogenesis. Although these factors appear not to be involved in the same high-risk populations, they do appear to be an important part of the etiology of esophageal cancer in the West. The roles of these factors in esophageal cancer are discussed in detail elsewhere in this volume.

IV. HOT DRINKS

Tea drinking is a common dietary feature of many of the regions that experience high esophageal cancer rates. This, combined with the concept that chronic thermal irritation of the esophageal mucosa may enhance the susceptibility of the tissue to carcinogenic stimuli, directs attention to the role of hot foods or drinks. While virtually all populations consume some portion of their diet cooked and served hot, carefully controlled measurement of food intake temperatures in populations is a difficult research task. The reporting of food habits by esophageal cancer cases may be affected by awareness of diagnosis or changes in dietary habits may have occurred as a result of the disease. Nevertheless, food and drink temperature intake often has been investigated and the research does offer some support for this etiologic hypothesis.

A case-control investigation of esophageal cancer among Singapore Chinese revealed statistically significant differences in reporting consumption of beverages at "burning hot" temperatures.[1] The presence of elevated adjusted relative risks ranging from 3- to 15-fold were observed for subjectively reported consumption of hot tea, coffee, and barley beverage among both males and females. Interestingly, the frequency of consumption of these beverages was not associated with greater risk. The association appeared to have a dose-response pattern and multivariate analyses failed to diminish the effect. Although the associations were strong and appeared not to have been confounded by other sources of risk, the investigators were careful to note that the temperatures of beverage intake were not measured directly.

Martinez found in a combined study of cancers of the esophagus, mouth, and pharynx that among males and, for both sexes combined, cases were significantly more likely to report consumption of hot coffee.[2] The tea temperatures of Turkomans, a high-risk group, were found to be significantly higher than for other groups in the Caspian region.[3] Two case-control studies of esophageal cancer in the Aktubisk province of Kazakhstan reported by Brunning[4] indicated more frequent consumption of hot tea by cases. Survey data from the high-risk region of Linxian province in China indicated that 77% of that population habitually consumed foods at high temperatures, but the absence of a relevant comparison population provides little indication of the etiologic importance of the observation.[5]

V. SILICA FIBERS

A recent development in the search for causative agents for esophageal cancer is the finding, in at least two regions of exceptional incidence, of high levels of mineral fibers as a contaminant of the diet. Flour and grain samples obtained from northeastern Iran revealed that bread consumed in that area is contaminated by fine silica fibers.[6] The fibers are comparable in size and shape to carcinogenic mineral fibers, e.g., large asbestos fibers. When mouse fibroblast cell cultures were treated with the fibers, a 100-fold increase in cell proliferation was observed. The same group of investigators subsequently has reported the presence of similar fibers in millet bran from northern China.[7] Their study was extended to include examination of mucosal specimens from esophageal cancer patients, at necropsy, of subjects in a low incidence area, London. The tissue from Linxian, China region revealed ten times the concentration of silica fibers compared to that of the London material.

These observations are significant because they may identify an exposure that is unique to two of the high-risk regions where the predominant risk factors, alcohol and tobacco, are in little evidence. The etiologic hypotheses relating to silica fibers are similar to those that have been posited for the carcinogenic action of asbestos. Those are, that the fibers lodge in the esophageal mucosa and provide anchorage for the collection of chemical carcinogens or that the fibers stimulate the proliferation of cells and increase the opportunities for a mutagenic or carcinogenic event. Much work remains to be accomplished regarding the role of silica fibers in esophageal carcinogenesis. Van Rensberg[8] has criticized published work on this topic from the standpoint of the inadequacy of the control populations that have been studied and has observed that, while millet bran is found in the Linxian Chinese diet, its consumption represents a minor aspect of the diet. Additional study of other populations exposed to plant mineral fibers to determine whether high incidence rates may be associated with this factor on a broader scale would be useful.

VI. DIETARY VITAMIN A

In recent years, several dietary components have come under scrutiny as possible inhibitors of carcinogenesis. Important among these is dietary vitamin A which is consumed in the diet in a preformed form such as retinol in animal fats or, in a precursor form, beta-carotene, in green and yellow vegetables. Vitamin A intake has been linked to reduced risk of a variety of tumor types in populations in modern, western societies. Peto et al.,[9] Kummet and colleagues,[10] and others have reviewed the studies and rationale for the association of vitamin A to cancer inhibition.

There are several different epidemiologic approaches to studying the role of dietary vitamin A in cancer inhibition in human populations. Population nutritional surveys can characterize the levels of intake of dietary vitamin A (and other nutrients) and the observed levels may be compared to some predesignated standard such as governmental standards of adequacy. It also is possible to measure level of retinol or carotene in the serum of subjects and to compare observed levels of those measured in some referent group. Finally, it is feasible to measure dietary intake of vitamin A by assessing the frequency of consumption of foods that are common sources of this nutrient. No single one of these approaches is completely adequate so it is difficult to draw conclusions from evidence based on but one approach. In light of this, it is interesting to note that the question of the association of esophageal cancer with dietary vitamin A has been studied by a number of approaches. However, as so often is the case when different methods are employed in different populations to address the same question, the aggregate evidence concerning the role of dietary vitamin A in inhibition of esophageal cancer is equivocal.

Groenewald et al.[11] conducted a survey of children and nursing mothers by means of the 24-hr recall method in Transkeians of the region of Africa with high esophageal cancer rates. Although one might question the appropriateness of the individuals surveyed to reflect esophageal cancer risk, the data do provide a description of some important dietary customs of the region. For example, maize is the dominant staple and pumpkins are the principal vegetable. Both are sources of vitamin A in the form of beta-carotene. Thus, in this population, although there were many nutritional inadequacies, the general pattern did not suggest that vitamin A deficiency would be prevalent. These observations are not entirely consistent with a study of the population of this region by van Rensberg et al.[12] by chemical analyses of blood specimens. They observed that the levels found in subjects from the highest risk region were greater than those for subjects from areas of more moderate esophageal cancer rates. Carotene levels, however, were significantly low and were lower in the population from the higher risk area, a finding that conflicts with the dietary survey data.

In Linxian, China, Yang et al.[13] studied serum levels and dietary habits in a group of subjects with a family history of esophageal cancer, a group with no family history, a group of subjects from another, lower risk region. No differences in serum retinol levels were evident between the different groups. While the levels observed generally were lower than observed in certain U.S. groups, only a small percentage of subjects were judged to be vitamin A deficient. Beta-carotene was not specifically assayed but the total carotenoid levels of the higher risk subjects were higher than for the lower risk groups.

Dietary surveys in Iran's exceptionally high-risk region and in comparison populations measured historical patterns of food consumption.[14] These data indicated low intake of fresh fruits and vegetables and, consequently, low intake of vitamin A. This observation was confirmed in a subsequent case-control study that further suggested that infrequent fruit and vegetable consumption was independent of socioeconomic status in its effect on esophageal cancer risk.[15]

Two case-control investigations in the U.S. have shown results similar to those observed in the Iran study. Ziegler et al.[16] and Mettlin et al.[17] both have reported that esophageal cancer patients report less frequent historical intake of fruits and vegetables that usually are sources of vitamin A in the diet. However, in both investigations, the authors were hesitant to attribute the difference in risk specifically to vitamin A since the foods studied were also sources of other important nutrients such as vitamin C. The levels of risk observed in our study of patients at Roswell Park are shown in Table 1. While significant levels of risk elevation were observable, additional analyses controlling for tobacco and alcohol use diminished the association to less than statistically significant levels.

An additional study in the Washington D.C. area examined plasma vitamin A levels of esophageal cancer patients relative to those of a control group of alcoholics and a group of elderly healthy men.[18] The cancer patients were observed to have significantly lower levels of plasma vitamin A compared to either group. The cancer patients were not otherwise observed to be malnourished but it remains possible that the low levels of plasma vitamin A were secondary to the disease rather than an antecedent to it. Acutal dietary intake was not measured in any of the populations studied and the relationship of plasma vitamin A to the possible dietary habits is not known.

VII. VITAMIN C

Mirvish[19] and others[20] have demonstrated that the formation of certain nitroso compounds linked in animal models to gastric carcinogenesis may be inhibited by the presence of vitamin C. The biological plausibility of the model and the observation that high-incidence regions and populations appear not to have continuous access to fresh fruits and vegetables that often are the main sources of vitamin A in the diet have led several to investigate vitamin C as an etiologic factor in esophageal cancer.

A major problem is assessing the effects of vitamin C, and most other nutrients, is that it does not occur in isolation in the diet. The foods that are sources of vitamin C often are also sources of vitamin A as well as other nutrients. Many of the epidemiologic findings linking vitamin A to esophageal cancer risk cited above could equally relate vitamin C to the observed risks. For example, the analysis of blood samples from groups in Linxian with

Table 1
RELATIVE RISKS FOR ESTIMATED
MONTHLY VITAMIN A INTAKE

Vitamin A[a]	Cases		Controls		
	OBS	EXP	OBS	EXP	RR
0—50 IU	38	32.46	57	62.54	1.28
50—100 IU	54	50.42	93	96.76	1.12
100+ IU	30	39.30	85	75.70	0.68
Total	122		235		

Note: Chi-square for trend = 4.67; p = 0.033.

[a] Calculated index of monthly vitamin A intake in thousands
  of International Units.

From Mettlin, C., Graham, S., Priore, R., Marshall, J., and
Swanson, M., *Nutr. Cancer,* 2, 143, 1981. With permission.

a family history of esophageal cancer, those without a family history, and a group from a
lower risk region showed that the family history group had lower levels of ascorbate.[12]
However, this was also shown to be true for vitamin A.

Infrequent consumption of fruits and vegetables has been reported in case-control studies
in Washington D.C.,[16] Buffalo, N.Y.,[17] and, the Caspian littoral of Iran.[15] The Buffalo
study is that carried out at Roswell Park Memorial Institute cited earlier. It was based on
the analysis of dietary data routinely obtained from patients entering the Institute; 147
esophageal cancer patients were compared to patients without cancer or other diseases of
the GI system. Overall fruit and vegetable consumption was associated with reduced risk;
persons reporting fruit and vegetable consumption 31 to 40 times a month had significantly
greater risk than those who reported consumption 81 times a month or more. The effect of
fruit and vegetable consumption was evident even after the effects of tobacco and alcohol
consumption were controlled. Interesting, with respect to the question of differentiating the
effects of vitamin A from vitamin C, was the observation that the relative risks for vitamin
A intake, as measured by a computed index of intake, were not significant after adjusting
for alcohol and smoking habits. In contrast, the risks for low vitamin C index, shown in
Table 2, remained significant. Those with the lowest level of the index were over two times
more likely to be esophageal cancer patients compared to those with the highest levels.

Tuyns,[21] in a study of 743 esophageal cancer patients in the Calvados region of France,
recently has reported findings that are quite consistent with the observations of the Roswell
park patient population. When compared to 1976 control subjects, esophageal cancer patients
were significantly less likely to report frequent consumption of citrus fruits, even when the
effects of alcohol consumption were considered. Table 3 presents the essential findings
showing that persons reporting low citrus fruit consumption and high levels of alcohol
consumption were nearly eight times more likely to be cases than controls. While alcohol
consumption appeared to be a more potent predictor of risk, even among light consumers
of alcohol, persons with low intake of citrus fruits were more likely to be esophageal cancer
cancers than controls. The same pattern of risk reduction for citrus fruit consumption was
observed among the heavier drinkers.

VIII. OTHER VITAMINS

As noted earlier, nutritional deficiencies, or relative deficiencies, seldom occur in isolation.
Because many foods are sources of multiple nutritional qualities, observation of one defi-

Table 2
RELATIVE RISKS FOR ESTIMATED VITAMIN C
INTAKE, STANDARDIZED BY SMOKING HABITS AND
ALCOHOL CONSUMPTION

Vitamin C intake	Cases		Controls		Smoking and alcohol adjusted (RR)
	OBS	EXP	OBS	EXP	
0—1000 mg	44	31.57	45	57.43	1.78
1001—1400 mg	27	33.16	75	68.84	0.75
1401+ mg	27	33.27	68	61.73	0.74
Total	98		188		

Note: Chi-square for trend = 8.94; p = 0.004

From Mettlin, C., Graham, S., Priore, R., Marshall, J., and Swanson, M., *Nutr. Cancer*, 2, 143, 1981. With permission.

Table 3
ESOPHAGEAL CANCER: RELATIVE RISK
IN EACH COMBINED VITAMIN C-
ALCOHOL CONSUMPTION GROUP

Group	Light consumers of alcohol ($\leq$40 g/day)	Heavy consumers of alcohol (>40 g/day)
Males		
Consumers of vitamin C	1.00	7.10
Nonconsumers of vitamin C	1.73	9.53
Females		
Consumers of vitamin C	1.00	6.62
Nonconsumers of vitamin C	1.76	13.28
Males and Females		
Consumers of vitamin C	1.00	5.90
Nonconsumers of vitamin C	1.75	7.92

From Tuyns, A. J., *Nutr. Cancer*, 5, 195, 1983. With permission.

ciency is likely to lead to observation of other dietary inadequacies. Thus, it is not surprising to find that in high esophageal cancer risk populations that have lower levels of vitamin A or C, owing to infrequent consumption of fresh fruits and vegetables, other vitamin deficiencies have been observed.

Nutrition surveys in Linxian, China, in the Transkei region of Africa, and in the high esophageal cancer risk populations of Iran all indicate lower levels of riboflavin in the diet compared to nutritional standards of adequacy or to referent populations.[5,11-14] A case-control study in Iran using a 24-hr dietary recall questionnaire showed that esophageal cancer patients had significantly lower levels of riboflavin intake than did controls.[22] In Linxian, analysis of sera from persons with family history of esophageal cancer revealed significantly lower levels of riboflavin activity compared to persons with no family history or to persons from another, lower risk, region.[13]

Rising incidence of pellegra has been reported to have preceded the onset of high rates of esophageal cancer in the Transkei.[23] Pellegra is a nutritional deficiency disorder common where corn is a dietary staple. Corn is deficient in nicotinic acid (niacin) and its precursor, tryptophan. Recent surveys in this high esophageal cancer risk population show that the local diets continue to be deficient in nicotinic acid.[24]

IX. MINERALS

Given as they are to appearance in remote regions, high-risk populations for esophageal cancer have evoked ingenious epidemiological techniques from those who would study its etiology. Few approaches are more ingenious than that employed by Burrell and associates[25] in the Transkei in the mid-1960s. They examined the characteristics of the gardens from which Bantu esophageal cancer victims had subsisted for no less than 15 years and the gardens of others from the Transkei with no esophageal cancer experience. The gardens of the esophageal cancer Bantu exhibited low productivity and plant morphological changes that suggested soil mineral deficiencies. While molybdenum deficiency was most evident, these gardens also were characterized by deficiencies in zinc, copper, iron, and other trace elements. Other interesting observations such as those relating to the gullet cancer experience of domestic fowl in China and esophageal lesions among cattle in the Transkei further suggest that soil, by virtue of its mineral content, may affect human esophageal cancer risk.[5,23]

Some laboratory data also implicate trace elements in esophageal cancer etiology. For example, Fong et al.[26] treated zinc-deficient rodents with a carcinogen, methylbenzylnitrosamine, and observed that compared to normal diet rats, the zinc-deficient rats experienced a significantly higher incidence of esophageal tumors. In human populations, van Rensberg has observed that populations of regions of high esophageal cancer incidence in the Transkei have diets deficient in dietary zinc as well as magnesium and calcium.[12] In Washington D.C., plasma zinc levels were found to be significantly lower among esophageal cancer patients compared to age-matched healthy controls.[18] In Iran, analysis of hair zinc levels from normal populations in the high incidence regions compared to samples obtained from a group in the U.S. and in Tehran suggested that the population of the high-risk region was relatively zinc deficient.[27] In high-risk regions of China, environmental studies as well as analyses of hair samples and urine have revealed relative deficiencies of molybdenum, zinc, manganese, and other trace elements.[28]

X. MUTAGENS AND CARCINOGENS

In two high-risk regions, Japan and China, preservation of vegetables by pickling is commonplace. Extracts from several pickled foods of the Linxian region of China have proven mutagenic in the Ames test.[29] These foods also have been analyzed and found to be high in content of nitroso compounds as well as containing fungi that promote the formation of nitrosamines. Thus, there is good reason to suspect that the manner of food preservation may play a significant role in the etiology of esophageal cancer. Consumption of moldy foods is common in high-risk regions and these molds also have mutagenic properties. Indeed, one case-control study in Linxian showed that esophageal cancer patients were nearly three times more likely than controls to report consumption of moldy cheese and moldy smoked meat.[30] Other evidence that food preservation practices may be related to esophageal cancer is the finding in Japan that household patterns of consumption of pickled vegetables correlated with regional patterns of esophageal cancer mortality.[31]

In addition to the widely observed association of alcohol consumption with esophageal cancer risk in the developed nations, there may be some alcohol drinking practices in certain regions that have particular effects on esophageal cancer risk. McGlashan[32] and Cook[33] have hypothesized that alcoholic drinks made with maize may relate to the pattern of risk in Africa. Drinks locally distilled from sugar and maize husks have been reported to be high in content of nitrosamines and the pattern of consumption of the alcoholic products are believed to relate to the pattern of risk in east Africa.

XI. DISCUSSION

This examination of the dietary risk factors for esophageal cancer has focused on putative sources of risk for which there are findings from multiple investigations from multiple research settings. Even avoiding consideration of isolated findings or uncontrolled observations, a range of possible dietary factors must be regarded as possibly playing some role in the etiology of esophageal cancer. They include:

1. Some dietary deficiency or deficiencies increase the susceptibility of the individual to the effects of carcinogenic stimuli such as tobacco
2. Some surfeit of nutritional intake provides protection against rare or ubiquitous environmental carcinogens
3. Some aspect of the diet itself constitutes a mutagenic or carcinogenic stimulus

Given the distribution of high-risk populations across cultures and regions that differ greatly in their dietary habits and environmental exposures, it seems likely that esophageal cancer has multiple etiologies. The evidence suggests that in western, developed areas, the risk-enhancing effects of tobacco and alcohol may be further enhanced by relatively low intakes of fresh frutis and vegetables. Findings from studies in Washington D.C., Buffalo N.Y., and in Calvados, France all are consistent with this general model of causation. However, in high-risk populations regions among whom tobacco and alcohol use are not common, a different model must be applied. In high-risk population regions among whom tobacco and alcohol use are not common, a different model must be applied. In high-risk areas such as China, Africa, and Iran, the carcinogenic stimuli may be represented by thermal irritation, trauma from mineral fibers, or carcinogens from molds or nitroso compounds in the diet. These sources of risk, combined with deficiencies in vitamin and or mineral intake may operate synergistically to generate the exceptionally high rates of the disease.

The complexity of the dietary epidemiology of esophageal cancer does not, at present, suggest that any single intervention to alter the dietary habits of populations at risk will be successful. In every setting studied, multiple factors appear possibly to operate. This may suggest that the means of risk reduction and prevention will not come from eliminating single exposures to risk. Another, more promising approach may be to identify a fundamental intervention that operates late in the disease etiology to disrupt the complex chain of events leading to cancer. Chemoprevention may be particularly suited to the problem of esophageal cancer. If it were possible to identify some agent such as selenium, a retinoid, vitamin C, or another vitamin which, when added to the diet, would counteract the effects of multiple carcinogens and deficiencies, a more feasible public health measure may result. This will ultimately require a shift in the research methodologies applied to the study of esophageal cancer. The observational study that has been the mainstay of epidemiological study of this topic may be replaced by the controlled prospective population intervention trial. Because therapies for esophageal cancer are so ineffective and because of epidemiologists' ability to identify specific populations of exceptionally high risk, this approach ultimately may represent the most effective strategy for control of esophageal cancer.

REFERENCES

1. **De Jong, U. W., Breslow, N., Goh Ewe Hong, J., Sridharan, M., and Shanmugaratnam, K.,** Aetiological factors in oesphageal cancer in Singapore Chinese, *Int. J. Cancer*, 13, 291, 1974.
2. **Martinez, I.,** Factors associated with cancer of the esophagus, mouth, and pharynx in Puerto Rico, *J. Natl. Cancer Inst.*, 42, 1069, 1969.
3. **Mahboubi, E. O., Aramesh, B.,** Epidemiology of esophageal cancer in Iran, with special reference to nutritional and cultural aspects, *Prev. Med.*, 9, 613, 1980.
4. **Brunning, D. A.,** Oesophageal cancer and hot tea, *Lancet*, 2, 272, 1974.
5. **Yang, C. S.,** Research on esophageal cancer in China: a review, *Cancer Res.*, 40, 2633, 1980.
6. **O'Neill, C. H., Hodges, G. M., Riddle, P. N., Jordan, P. W., Newman, R. H., Flood, R. J., and Toulson, E. C.,** A fine fibrous silica contaminant of flour in the high oesphageal cancer area of north-east Iran, *Int. J. Cancer*, 26, 617, 1980.
7. **O'Neill, C., Clarke, G., Hodges, G., Jordan, P., Newman, R., Pan, Q., Liu, F., Ge, M. M., Chang, Y., and Toulson, R.,** Silica fragments from millet bran in mucosa surrounding oesophageal tumours in patients in northern China, *Lancet*, 1, 1202, 1982.
8. **Van Rensberg, S. J.,** Oesophageal cancer, micronutrient malnutrition, and silica fragments, *Lancet*, 2, 1098, 1982.
9. **Peto, R., Doll, R., Buckley, J. D., and Sporn, M. B.,** Can dietary beta-carotene materially reduce human cancer rates?, *Nature (London)*, 290, 201, 1981.
10. **Kummet, T., Moon, T. E., and Meyskens, F. L.,** Vitamin A: evidence for its preventive role in human cancer, *Nutr. Cancer*, 5, 96, 1983.
11. **Groenewald, G., Langenhoven, M. L., Beyers, M. J. C., Du Plessis, J. P., Ferreira, J. J., and van Rensberg, S. J.,** Nutrient intakes among rural Transkeians at risk for oesophageal cancer, *S. Afr. Med. J.*, 60, 964, 1981.
12. **Van Rensberg, S. J., Benade, A. S., Rose, E. F., and du Plessis, J. P.,** Nutritional status of African populations predisposed to esophageal cancer, *Nutr. Cancer*, 4, 206, 1093.
13. **Yang, C. S., Miao, J., Yang, W., Huang, M., Wang, T., Xue, H., You, S., Lu, J., and Wu, J.,** Diet and nutrition of the high esophageal cancer risk population in Linxian, China, *Nutr. Cancer*, 4, 154, 1982.
14. **Hormozdiari, H., Day, N. E., Aramesh, B., and Mahboubi, E.,** Dietary factors and esophageal cancer in the Caspian littoral of Iran, *Cancer Res.*, 35, 3493, 1975.
15. **Cook-Mozaffari, P. J., Azordegan, F., Day, N. E., Ressicaud, A., Sabai, C., and Aramesh, B.,** Oesophageal cancer studies in the Caspian littoral of Iran: results of a case control study, *Br. J. Cancer*, 39, 293, 1979.
16. **Ziegler, R. G., Morris, L. E., Blot, W. J., Pottern, L. M., Hoover, R., and Fraumeni, J. F.,** Esophageal cancer among black men in Washington D.C. II. Role of nutrition, *J. Natl. Cancer Inst.*, 67, 1199, 1981.
17. **Mettlin, C., Graham, S., Priore, R., Marshall, J., and Swanson, M.,** Diet and cancer of the esophagus, *Nutr. Cancer*, 2, 143, 1981.
18. **Mellow, M. H., Layne, E. A., Lipman, T. O., Kaushik, M., Hostetler, C., and Smith, J. C.,** Plasma zinc and vitamin A in human squamous carcinoma of the esophagus, *Cancer*, 51, 1615, 1983.
19. **Mirvish, S. S.,** Inhibition of the formation of carcinogenic N-nitroso compounds by ascoric acid and other compounds, in *Cancer: Achievements, Challenges, and Propsects for the 1980s*, Vol. 1, Burchenal, J. H. and Oettgen, H. F., Eds., Grune & Stratton, New York, 1981, 557.
20. **Weisburger, J. H., Wynder, E. L., and Horn, C. L.,** Nutritional factors and etiologic mechanisms in the the causation of gastrointestinal cancers, *Cancer*, 50, 2541, 1982.
21. **Tuyns, A. J.,** protective effect of citrus fruit on esophageal cancer, *Nutr. Cancer*, 5, 195, 1983.
22. **Siassi, F.,** Riboflavin and oesophageal cancer, *Fed. Proc.*, 3, 654, 1980.
23. **Warwick, G. P. and Harington, J. S.,** Some aspects of the epidemiology and etiology of esophageal cancer with particular emphasis on the Transkei, South Africa, *Adv. Cancer Res.*, 17, 82, 1973.
24. **Van Rensberg, S. J.,** Epidemiologic and dietary evidence for a specific nutritional predisposition to esophageal cancer, *J. Natl. Cancer Inst.*, 67, 243, 1981.

25. **Burrell, R. J. W., Roach, W. A., and Shadwell, A.,** Esophageal cancer of the Transkei associated with mineral deficiency in garden plants, *J. Natl. Cancer Inst.,* 36, 201, 1966.
26. **Fong, L. Y. Y., Sival, A., and Newberne, P. M.,** Zinc deficiency and methylbenzylnitrosamine-induced esophageal cancer in rats, *J. Natl. Cancer Inst.,* 61, 145, 1978.
27. **Mobarhan, S., Dowlatshahi, K., and Diba, Y. Y.,** Hair zinc levels from a normal population of north east Iran with a high incidence of esophageal cancer, *Am. J. Clin. Nutr.,* 33, 940, 1980.
28. **Li, J. Y.,** Epidemiology of esophageal cancer in China, *Natl. Cancer Inst. Monogr.,* 62, 113, 1982.
29. **Mingxin, L., Ping, L., and Baorong, L.,** Recent progress in research on esophageal cancer in China, *Adv. Cancer Res.,* 33, 173, 1980.
30. **Lu, S., Camus, A., Tomatis, L., and Bartsch, H.,** Mutagenicity of extracts of pickled vegetables collected in Linhsien County, a high-incidence area for esophageal cancer in northern China, *J. Natl. Cancer Inst.,* 66, 33, 1981.
31. **Nagai, M., Hashimoto, T., Yanagawa, H., Yokoyama, H., and Minowa, M.,** Relationship of diet to the incidence of esophageal and stomach cancer in Japan, *Nutr. Cancer,* 3, 257, 1982.
32. **McGlashan, N. D.,** Cancer of the oesophagus and alcoholic spirits in central Africa, *Gut,* 10, 643, 1969.
33. **Cook, P.,** Cancer of the oesophagus in Africa, *Br. J. Cancer,* 24, 853, 1970.

Chapter 6

CANCER OF THE ESOPHAGUS: EPIDEMIOLOGICAL AND EXPERIMENTAL STUDIES

G. David McCoy

TABLE OF CONTENTS

I. INTRODUCTION

Esophageal cancer is one of a family of cancers which can be grouped under the heading of cancers of the head and neck. These cancers which include oral, esophageal, and laryngeal have several common features besides their anatomical location which strongly suggest that they may be etiologically related. Beginning with the classic studies of Wynder and associates, almost 30 years of retrospective and propsective epidemiological studies have clearly demonstrated that both the incidence of and mortality from cancers of the head and neck in the U.S. and other industrialized nations, are profoundly increased in persons, particularly males, who are heavy users of both tobacco and alcohol.[1-15]

The American Cancer Society figures for 1983 estimate that there will have been 47,100 new cases and 21,450 deaths in 1983 from cancers of the head and neck. Esophageal cancer comprises 19% of the 1983 estimated cases of head and neck cancer but accounts for almost 40% of the deaths in this group.[16] Five year survival rates have remained at 4% from 1950 to 1973.[16,17] Although cancers of the head and neck account for only approximately 5% of all new cancer cases and deaths, the certain association of heavy alcohol and tobacco use with increased risk for cancers at these sites ranks them among the most preventable of all cancers in man.

Rates of esophageal cancer incidence and mortality are two to three times higher amongst blacks than whites in the U.S.[18] Analysis of age-specific incidence curves indicates that esophageal cancer occurs at an earlier age in blacks than in whites but that the incidence curves tend to converge beyond 70 years of age.[19] The reasons for racial differences in esophageal cancer are not clear. Del Villano et al. have shown that black male alcoholics have elevated red blood cell superoxide dismutase (E.C.1.15.1.1) compared to both alcoholic white and nonalcoholic black males.[20] The relationship of this observation if any to increased risk for esophageal cancer is not known. Estimation of activities such as superoxide dismutase may prove to be useful screening tests for persons at risk for esophageal cancer and other alcohol-related diseases. Keller has noted that in terms of the total amount of alcohol consumed, black and white male cancer patients consume the same total amount of ethanol; however, distilled spirit consumption was significantly higher in blacks than whites.[21] These data as well as the earlier onset of esophageal cancer in blacks suggests that substances other than ethanol in distilled spirits may be exerting a promotional effect on esophageal cancer.

II. OCCURRENCE AND POSSIBLE INVOLVEMENT OF BEVERAGE CONGENERS

In addition to ethanol and water, nearly all alcoholic beverages contain variable amounts of long chain alcohols, acids, esters, carbohydrates, traces of protein, peptides and free amino acids, vitamins, minerals, and other compounds peculiar to the individual beverage.[22,23] Kahn has cataloged nearly 400 organic compounds that can be found in alcoholic beverages.[24] The group of long chain alcohols collectively known as fusel oils are composed mainly of 1-propanol, 2-methyl-1-propanol, 3-methyl-1-butanol, and 2-methyl-1-butanol although longer aliphatic alcohols have been demonstrated.[25] These alcohols are by-products of the fermentation process and arise from the sequential transamination decarboxylation and reduction of endogenous amino acids.[26] Administration of these compounds to rats has shown them to be both hepatotoxic and carcinogenic.[27,28]

The long chain alcohols including the fusel oils have been shown to be substrates for alcohol dehydrogenase (E.C.1.1.1.1.) from human, horse, rat, and hamster liver.[29,30] Extra hepatic alcohol dehydrogenase activity has also been demonstrated in epithelial preparations from hamster cheek pouch epithelium[30] and rat stomach.[31] Preparations from hamster cheek pouch metabolized certain of the fusel oils at rates greater than those obtained when ethanol served as substrate.[30] These data indicate that both ethanol and fusel oil metabolism can occur at sites other than the liver and that ethanol and/or other alcohols can exert direct effects on intracellular metabolism at sites other than the liver. The aldehydes which result from metabolism of ethanol and the longer chain alcohols are highly reactive species and have the potential of exerting either toxic or mutagenic effects. Acetaldehyde has been shown to be a mutagen for Ames strain TA102[32] and also to induce sister chromatid exchange in cultured Chinese hamster ovary cells.[33,34] Unfortunately no studies on the ability of extracts from human esophageal or oral epithelial preparations to metabolize alcohols have been reported.

Much interest has been focused on the presence of carcinogens and mutagens in alcoholic beverages, their extracts, and condensates using the Ames *Salmonella* test system test as an indicator of mutagenic activity.[38] These studies indicate that mutagenic activity can be demonstrated in alcoholic beverages and in general, the mutagenic activity of home-manufactured beverages is higher than their commercially manufactured counterparts.[36]

In 1966 Masuda et al. reported the presence of numerous polycyclic aromatic hydrocarbons including known carcinogens in eight brands of scotch, four brands of bourbon, and two brands of Japanese whiskey.[39] Subsequent studies by this group failed to demonstrate any carcinogenic activity when residues of these whiskeys were applied to mouse skin. However, some promoter-like activity was observed in a Japanese whiskey sample.[40] With the advent of highly sensitive detectors which can specifically detect nitrosamines, several laboratories have examined various types of alcoholic beverages for the presence of nitrosamines. Spiegelholder et al. demonstrated the presence of dimethylnitrosamine (DMN) in 111 of 158 samples of various types of German beer tested.[41] DMN levels were on average 2.7 pbb with the highest reported level of 68 ppb. Similar levels of DMN have been observed in the beers produced in the U.S.[42] Goff and Fine reported similar levels in domestic and imported beer as well as all seven samples of scotch whiskey tested. No detectable levels were found in other types of distilled or fermented beverages.[43] It should be pointed out that although the detected levels of DMN are extremely low, in view of the known animal carcinogenicity of the *N*-nitrosamines as well as their ability to induce esophageal tumors in experimental animals,[44,45] the efforts of the brewers and distillers to remove these compounds from their products is to be commended.

III. ROLE OF TOBACCO-ASSOCIATED CARCINOGENS

Tobacco and tobacco smoke contain many distinct classes of chemical carcinogens and co-carcinogens. The polynuclear aromatic hydrocarbon (PAH) fraction of tobacco smoke is a major contributor to the carcinogenic activity in smoke. Benzo(*a*)pyrene (B(*a*)P) is the major unsubstituted PAH found in tobacco smoke. Levels of B(*a*)P in smoke have been shown to be good indicators of the tumorigenicity of smoke on mouse skin, as well as in inhalation experiments with Syrian golden hamsters.[46] The PAH in general and B(*a*)P specifically are known as contact carcinogens in that they produce tumors at the site of application.[47] These compounds must be metabolized intracellularly to their ultimate carcinogenic forms. This process which is called metabolic activation has been shown to be catalyzed by the microsomal cytochrome P_{450} associated mixed function oxygenase system. Innumerable studies have established that metabolic activation of B(*a*)P can occur in virtually any tissue which contains the cytochrome P_{450} system.[48]

A second major class of carcinogens found in tobacco and tobacco smoke are the *N*-nitrosamines. The nitrosamines, *N*-nitrosopyrrolidine (NPYR), N′-nitrosonornicotine (NNN), 4-(*N*-methyl-*N*-nitrosamino)-1-(3-pyridyl)-1-butanone (NNK), and dimethylnitrosamine (DMN) are important environmental carcinogens. NPYR has been detected in bacon and other process meats, as well as in mainstream and sidestream tobacco smoke.[49,50] NNN and

NNK are the major tobacco-specific carcinogens, occurring in tobacco and tobacco smoke in relatively high concentrations.[51] Oral administration of NPYR gives liver tumors in rats[52] and lung adenomas in strain A mice.[53] Oral administration of NNN causes esophageal tumors in rats while s.c. injection gives nasal cavity tumors.[54,55] S.c. administration to hamsters gives nasal cavity and tracheal tumors.[56] NNK causes lung, nasal cavity, and tracheal tumors in hamsters.[57]

α-Hydroxylation as a likely activation mechanism for nitrosamines was first suggested by Druckrey.[44] The role of α-hydroxylation in the activation of nitrosamines, such as NPYR and NNN, has been supported by structure-activity studies in which decreased carcinogenicity was observed when methyl groups or deuterium atoms were substituted in the α-positions[58,59] and by the high mutagenicity, without activation of α-acetoxy cyclic nitrosamines and related model compounds.[60,61] The α-hydroxylation of N-nitrosamines has been shown to be catalyzed by the microsomal cytochrome P_{450} associated mixed function oxygenase system.[62-64]

IV. METABOLIC STUDIES WITH TOBACCO-ASSOCIATED CARCINOGENS

Harris et al.[65] have demonstrated that B(a)P, DMN, and NPYR are metabolized by cultured explants of human esophagus. Autrup and Stoner have shown that N-nitrosobenzylmethylamine (BMNA), diethylnitrosamine (DEN), N-nitrosomethylethylamine (EMN), DMN, and NPYR are metabolized by cultured human and rat esophagus.[66] Recently Hecht et al. have shown that the tobacco-specific carcinogen NNN is metabolized by cultured rat esophagus.[67] Because the two α carbons of NNN are not equivalent, α-hydroxylation results in two products, a keto-alcohol from 2'-hydroxylation and a lactol from 5'-hydroxylation. The authors point out that in liver the major site of α-hydroxylation is the 5' position. In the esophagus it is the 2' position which is more readily hydroxylated. Although there are now numerous examples of metabolism of carcinogens by esophageal preparations, not all esophageal carcinogens can be shown to be metabolized by esophageal preparations.[68,69]

V. ETHANOL AND EXPERIMENTAL CARCINOGENESIS

Beginning with the investigations by Protzel et al.,[70] several studies have shown that ethanol consumption can increase the tumor incidence in animals exposed to B(a)P[71,72] or 7,12-dimethylbenz(a)anthracene.[73-75] The role that ethanol plays in increased tumor incidence of PAH-treated animals is not known. However, the enhanced 3-hydroxylation and mutagenicity of B(a)P by subcellular fractions of small intestinal mucosa from ethanol-consuming rats indicates that ethanol can increase microsomal metabolism in at least one extra hepatic tissue.[76]

Several studies on the effect of alcohol on the carcinogenicity of nitrosamines have been reported. In 1965, Gibel[77] presented evidence that simultaneous administration of ethanol and diethylnitrosamine by gastric intubation increased the incidence of esophageal tumors in rats. Both control and ethanol-treated rats had 100% incidence of liver tumors. Animals treated with dinitrosopiperazine showed no increase due to ethanol. Schmahl[78] has shown that the carcinogenicity of N-nitrosomethylphenylamine administered either p.o. or s.c. to male and female Sprague-Dawley rats was not increased by ethanol consumption. Girciute et al. have shown simultaneous gastric intubation of ethanol and N-nitrosodi-n-propylamine (NDPA) to mice resulted in the increased number of animals with tumors, the number of tumors per animal as well as the number of malignant tumors compared to mice only receiving NDPA.[79] Our own studies have shown that oral administration of ethanol followed by i.p. injection of N-nitrosopyrrolidine (NPYR) increased both the number of animals with nasal cavity and tracheal tumors as well as the number of malignant tumors.[80] It should be noted that no increase in tumor incidence due solely ethanol was observed in any of the above studies. A study by Schrauzer et al.[81] has shown, however, that chronic ethanol consumption can decrease the latent period and increase the tumor volume of spontaneously occurring mammary adenocarcinoma in female C3H/St mice.

Although alcohol consumption in and of itself has been suggested as increasing the risk for esophageal cancer in the absence of tobacco smoking, it should be noted that few heavy drinkers are nonsmokers making it difficult to clearly establish the role of alcohol by traditional epidemiological means.[82]

VI. DIETARY FACTORS ASSOCIATED WITH CANCER OF THE ESOPHAGUS

In other parts of the world epidemiological studies suggest that alcohol and tobacco consumption are not significant risk factors for esophageal cancer. The possiblity that nutrition might play a role in the etiology of esophageal cancer was first suggested by consideration of the association of Plummer-Vinson (Paterson-Kelly) syndrome and esophageal cancer. This disease, which was once prevalent among Swedish women, was shown to be associated with chronic iron and vitamin deficiencies. High rates of upper alimentary tract cancer were observed in the absence of exposure to tobacco or any other obvious source of carcinogen.[83,84] Since the introduction of a national program of dietary iron and vitamin supplementation in Sweden in the early 1950s, a significant reduction of upper alimentary tract cancer.[85] In many areas where rates of esophageal cancer are the highest the occurrence of the disease in males and females approaches unity.[86] Thus, for example, Hormozdiari et al.[87] and Mahboubi and Aramesh[88] have shown that in the area of the Capian Litteral, located in north central Iran, male-female incidence rates approach unity in areas of moderate to high esophageal cancer incidence. The people in areas of moderate to high esophageal cancer incidence consume more wheat than rice, more sheep and goat milk and milk products, fewer fruits and vegetables and thereby lower intakes of vitamins A, B_2, and C. Li et al.[89] and Yang[90] have reviewed the literature pertaining to the epidemiology and etiology of esophageal cancer in the People's Republic of China.[89] In those areas of China where esophageal cancer rates are among the highest in the world the male-female ratios are close to unity. In these areas of high esophageal cancer decreased intake of vitamins A, B_2, and C, fresh fruits and vegetables, and protein and fat was found. Because deaths due to esophageal cancer represent 22% of all cancer deaths in the People's Republic of China this nation's health care policy has been directed toward the elimination of esophageal cancer. Their strategy is a four-pronged attack on the problem which includes studies of the factors in the environment which are associated with areas of high esophageal cancer. Second, increased surveillance in high-risk areas with emphasis on early diagnosis and treatment not only of the disease but of preneoplastic conditions. Large-scale introduction of preventative measures designed to reduce or eliminate these factors in the environment such as mold contamination of foods, reduction of nitrates, nitrites, and amines from drinking water, and introduction of molybdenum-enriched fertilizers which yield crops lower in nitrate and nitrite and higher in vitamin C.

Mellow et al.[91] have presented evidence to show that esophageal cancer patients have lowered plasma zinc and vitamin A levels. Graham[92] has noted an inverse relation between risk for esophageal cancer and vitamin A and C consumption. When he controlled for tobacco and alcohol consumption the association of vitamin A disappeared while the association with vitamin C remained. The demonstration by Meade et al.[93] that serum and liver vitamin A stores are lower in hamsters chronically exposed to tobacco smoke is consistent with the view that lowered levels of this vitamin may be associated with increased risk for cancer. Lee and Lieber[94] have shown that hepatic levels of vitamin A are lowered in both experimental animals and man as a consequence of alcohol consumption and that these changes precede changes in serum vitamin A levels.

Nutritional deficiencies are commonly associated with alcoholism.[95] Since alcoholics often consume 900 or more calories a day from alcohol alone,[96] it is not difficult to imagine that the rest of their dietary intake is insufficient in providing essential micronutrients. Alcohol consumption can lead to impaired absorption of nutrient and vitamins.[97] Thus, evidence suggests that nutritional deficiencies arising from either undernutrition and/or as a direct consequence of alcohol intake (impaired absorption or enhanced elimination) probably play a role in the etiology of esophageal cancer.

VII. ROLE OF DIET IN EXPERIMENTAL CARCINOGENESIS

Feeding mice a diet deficient in riboflavin causes morphological alterations in skin and upper alimentary tract epithelium which are similar to those observed in patients suffering from Plummer-Vinson disease.[98] As the deficiency progresses, epithelial morphology progressively changes from atrophy to hyperkeratosis to, in several instances, hyperplasia. The experiments of Chan and Wynder[99] have shown that, following initiation with B(a)P and promotion with croton oil, riboflavin-deficient mice develop tumors more rapidly than do control mice receiving a nutritionally adequate diet. In parallel studies, Chan et al.[100] have shown that basal levels of skin arylhydrocarbon hydroxylase were reduced slightly in riboflavin-deficient mice. However, the skin activity of riboflavin-deficient animals was induced to a much greater extent following a single application of dimethylbenz(a)anthracene (DMBA).

The work of Gerson and Meyer[101] has shown that feeding rats diets deficient in zinc causes similar changes in the morphology of the buccal mucosa.

Dietary zinc deficiency has been shown to increase the number of esophageal tumors and to decrease the latent period in rats exposed to methylbenzylnitrosamine.[102] Decreased levels of zinc have been observed in hair and tissue samples from esophageal cancer patients.[103] In animals made deficient for vitamin A, the tracheobronchial epithelium undergoes atrophic degenerative changes[104,105] quite similar to the changes observed in animals exposed to tobacco smoke.[106,107] Vitamin A-deficient animals have been shown to be more susceptible to PAHs or PAH carcinogenesis.[108]

VIII. STUDIES ON *IN SITU* NITROSATION

The realization that nitrosation of endogenous amines to form *N*-nitrosamines was possible under physiological conditions plus the fact that not only were the precursers such as nitrite and secondary amines present in the diets of persons in higher esophageal cancer areas but also that vitamin C intake was decreased in these same areas (vitamin C is a potent inhibitor of N-nitrosation), provided a strong impetus for screening high-risk areas for both nitrosamines and nitrosamine precursors. The National Academy of Sciences has recently published a monograph[109] entitled ''The Health Effects of Nitrate, Nitrite, and *N*-Nitroso Compounds'' which provides an outstanding review of the environmental exposure to these compounds as well as the chemistry of nitrosation and possible health effects.

Although numerous studies have indicated that the precursors necessary for *N*-nitrosamine formation are present in areas of high esophageal cancer, it has only been since 1980 that any systematic attempt has been made at measuring *in situ* nitrosation. Advantage is being taken of the fact that the amino acid proline can be readily nitrosated to form *N*-nitrosoproline (NPro). This compound has been shown to be noncarcinogenic in experimental animals[110,111] and is not metabolized to any significant extent.[112] The work of Ohshima and Bartsch[113] demonstrated that NPro is found in the urine following ingestion of proline and beet juice as the source of nitrate. Simultaneous ingestion of vitamin C prevented the appearance of NPro. Hoffmann and Brunnerman[114] examined the effect of cigarette smoking on the appearance of nitrosoproline. Although much variability was encountered their data indicate that cigarette smoking increased the appearance of NPro in urine and that ascorbate blocked its appearance.

IX. CONCLUSIONS

Epidemiological studies have clearly demonstrated a strong positive association of chronic alcohol and tobacco consumption with esophageal cancer. Though not as clear-cut, nevertheless diets deficient in specific vitamins (C, B₂, A) are also associated with increased risk for esophageal cancer. Simplistically it is tempting but probably premature to suggest that chronic alcohol consumption results in specific vitamin deficiencies thus forging a link between esophageal cancer incidence in westernized and third world nations. Since ethanol can interact with tissues and organ systems in many ways and can influence intracellular metabolism at many levels, it is probably an oversimplication to ascribe its association with esophageal cancer as simply that of a vitamin depleter. Ethanol administration to experimental animals has clearly been shown to increase tumor incidence of B(a)P as well as several N-nitrosamines with no increased tumor incidence in noncarcinogen ethanol-treated controls. The temporal relationship between alcohol consumption and carcinogen administration has not yet been established. In other words, which is the important time for exposure to alcohol to cause increased carcinogenicity? While it is true that in man, alcohol and tobacco consumption in persons at risk for esophageal cancer occur together over long periods of time, experimental studies should be able to determine whether alcohol consumption before, during, or after carcinogen exposure is important for increased tumor incidence. An understanding of the sequence of events resulting in increased carcinogenicity in experimental animals will help to pinpoint the important step in the series of steps which lead from normal cells to cancer cells which must be blocked to prevent neoplastic progression.

There are still large gaps in our knowledge of the basic biochemistry of the esophagus as well as the other extrahepatic tissues in both experimental animals and man. It is clear that metabolism and metabolic activation of chemical carcinogens can and does occur in these tissues. The effect of modifiers such as ethanol, dietary deficiencies, and/or excesses has not been studied in any detail. This lack of knowledge must be rectified if we are to understand the role of risk factors in disease etiology. Effective preventative measures need not wait for a thorough understanding of all factors which result in increased risk of esophageal cancer. An emphasis now on education of the public, espeicaly the young, as to all of the health consequences of alcohol and tobacco use can and should be done now.

REFERENCES

1. **Wynder, E. L., Bross, I. J., and Day, E. A.,** A study of environmental factors in the cancer of the larynx, *Cancer (Philadelphia),* 9, 86, 1956.
2. **Wynder, E. L. and Bross, I. J.,** Aetiological factors in mouth cancer, *Br. Med. J.,* 1, 1137, 1957.
3. **Wynder, E. L., Bross, I. J., and Feldman, R. M.,** A study of the etiological factors in cancer of the mouth, *Cancer (Philadelphia),* 10, 1300, 1957.
4. **Vincent, R. G. and Marchetta, F.,** The relationship of the use of tobacco and alcohol to cancer of the oral cavity, pharynx or larynx, *Am. J. Surg.,* 106, 501, 1063.
5. **Kamionkowski, M. D. and Fleshier, B.,** The role of alcoholic intake in esophageal carcinoma, *Am. J. Med. Sci.,* 249, 696, 1965.
6. **Keller, A. Z. and Ferris, M.,** The association of alcohol and tobacco with cancer of the mouth and pharynx, *Am. J. Public Health,* 55, 1578, 1965.
7. **Moore, C.,** Cigarette smoking and cancer of the mouth, pharynx and larynx, *JAMA,* 191, 104, 1965.

8. **Rothman, E. and Keller, A. Z.**, The effect of joint exposure to alcohol and tobacco on risk of cancer of the mouth and pharynx, *J. Chronic Dis.*, 25, 711, 1972.

9. **Wynder, E. L. and Mabuchi, K.**, Etiological and environmental factors in esophageal cancer, *JAMA*, 226, 1546, 1973.

10. **Schottenfeld, D., Gantt, R. C., and Wynder, E. L.**, The role of alcohol and tobacco in multiple primary cancers of the upper digestive system, larynx and lungs — a propsective study, *Prev. Med.*, 3, 277, 1974.

11. **Wynder, E. L., Covey, L. S., Mabuchi, K., and Mushinski, M.**, Environmental factors in cancer of the larynx, *Cancer (Philadelphia)*, 38, 1591, 1976.

12. **Graham, S., Dayal, H., Roher, T., Swanson, M., Sultz, H., Shedd, D., and Fischman, S.**, Dentition, diet, tobacco and alcohol in the epidemiology of oral cancer, *J. Natl. Cancer Inst.*, 59, 1611, 1977.

13. **Wynder, E. L. and Stellman, S. D.**, Comparative epidemiology of tobacco-related cancers, *Cancer Res.*, 37, 4608, 1977.

14. **Williams, R. R. and Horn, J. W.**, Association of cancer sites with tobacco and alcohol consumption and socioeconomic status of patients. Interview Study from the Third National Cancer Survey, *J. Natl. Cancer Inst.*, 58, 525, 1977.

15. **Schmidt, W. and Popham, R. E.**, Role of drinking and smoking in mortality from cancer and other cases in male alcoholics, *Cancer*, 47, 1031, 1981.

16. American Cancer Society, *Cancer Facts and Figures 1983*, American Cancer Society, New York, 1983.

17. **Axtell, L. M., Cutler, S. J., and Meyers, M. H., Eds.**, End Results in Cancer Report No. 4, DHEW Publication No(NIH) 73-272, Department of Health, Education and Welfare, Washington, D.C., 1972, 43.

18. **Keller, A. Z.**, The epidemiology of esophageal cancer in the west, *Prev. Med.*, 9, 607, 1980.

19. **Devesa, S. S.**, Reliability of reported death rates and incidence rates, *Prev. Med.*, 9, 589, 1980.

20. **Del Villano, B. C., Miller, S. I., Schacter, L. P., and Tischfield, J. A.**, Elevated superoxide dismutase in black alcoholics, *Science*, 207, 991, 1980.

21. **Keller, A. Z.**, Liver cirrhosis, tobacco, alcohol and cancer among blacks, *J. Natl. Med. Assoc.*, 70, 575, 1978.

22. **Leake, C. D. and Silverman, M.**, The chemistry of alcoholic beverages, in *The Biology of Alcholism*, Vol. 1, Kissin, B. and Begleiter, H., Eds., Plenum Press, New York, 1979, chap. 17.

23. **Greizerstein, H. B.**, Congener contents of alcoholic beverages, *J. Stud. Alchol*, 42, 1030, 1981.

24. **Kahn, J. H.**, Compounds identified in whiskey, wine and beer: a tabulation, *J. Am. Off. Anal. Chem.*, 52, 1166, 1969.

25. **Singh, R. and Kunkee, R. E.**, Alcohol dehydrogenase activities of wine yeast in relation to higher alcohol formation, *Appl. Environ. Microbiol.*, 32, 666, 1976.

26. **Sentheshanmuganathan, S.**, The mechanism of formation of higher alcohols from amino acids by *Saccharomyces cerevisiae*, *Biochem. J.*, 74, 568, 1960.

27. **Gibel, W., Wildner, G. P., and Lohs, Kh.**, Untersuchungen zur Prage einer Kanzerogenen und hepatotoxischen Wirkung von Fuselol, *Arch. Geschwulstforsch.*, 32, 115, 1968.

28. **Gibel, V. W., Lohs, Kh., Wildner, G. P., Wittbrodt, S., Geibler, E., and Hilscher, H.**, Untersuchungen zur Frage einer moglichen imtagenen Wirkung von Vuselol, *Arch. Geschwulstforsch.*, 33, 49, 1969.

29. **Petruszko, R.**, Nonethanol substrates of alcohol dehydrogenase, in *Biochemistry and Pharmacology of Ethanol*, Vol. 1, Majchrowicz, E. and Nobel, E. P., Eds., Plenum Press, New York, 1979, chap. 5.

30. **McCoy, G. D., Tambane, P. C., Powchik, P., and Teague, C. A.**, Occurrence and distribution of ethanol metabolizing enzymes in hamster cheek pouch epithelium, *Cancer Biochem. Biophys.*, 4, 111, 1981.

31. **Winer, A. D.**, Gastric alcohol dehydrogenase: a zinc metalloenzyme, *Fed. Proc.*, 37, 1967.

32. **Levin, D. E., Hollstein, M., Christman, M. F., Schwiers, E. A., and Ames, B. N.**, A new Salmonella tester strain (TA102) with A·T base pairs at the site of mutation detects oxidative mutagens, *Proc. Natl. Acad. Sci. U.S.A.*, 79, 7445, 1982.

33. **Obe, G., and Ristow, H.**, Acetaldehyde but not ethanol induces sister chromatid exchange in Chinese hamster cells in vitro, *Mutat. Res.*, 56, 211, 1977.

34. **Obe, G. and Beek, B.**, Mutagenic activity of aldehydes, *Drug Alcohol Dependence*, 4, 91, 1979.

35. **Lee, J. S. K. and Fong, L. Y. Y.**, Mutagenicity of Chinese alcoholic spirits, *Food Cosmet. Toxicol.*, 17, 575, 1979.

36. **Loquet, C., Toussaint, G., and LeTalaer, J. Y.**, Studies on the mutagenic constituents of apple brandy and various alcoholic beverages collected in western France, a high incidence area for esophageal cancer, *Mutat. Res.*, 88, 155, 1981.

37. **Nagao, M., Takahashi, Y., Wakabayashi, K., and Sugimura, T.**, Mutagenicity of alcoholic beverages, *Mutat. Res.*, 88, 147, 1981.

38. **Ames, B. N., McCann, J., and Yamasaki, E.**, Method for detecting carcinogens and mutagens with the Salmonella/mammalian-microsome mutagenicity test, *Mutat. Res.*, 31, 347, 1975.

39. **Masuda, Y., Mori, K., Hirohata, T., and Kuratsune, M.,** Carcinogenesis in the esophagus. III. Polycyclic aromatic hydrocarbons and phenols in whiskey, *Gann,* 57, 549, 1966.
40. **Kuratsune, M., Kohchi, S., and Horie, A.,** Test of alcoholic beverages and ethanol solutions for carcinogenicity and tumor promoting activity, *Gann,* 62, 395, 1971.
41. **Spiegelhalder, B., Eisenbrand, G., and Preussman, R.,** Contamination of beer with trace quantities of N-nitrosodimethylamine, *Food Cosmet. Toxicol.,* 17, 29, 1979.
42. **Scanlan, R. A., Barbour, J. F., Hotchkiss, J. H., and Libben, L. M.,** N-Nitrosodimethylamine in beer, *Food Cosmet. Toxicol.,* 18, 27, 1980.
43. **Goff, E. U. and Fine, D. H.,** Analysis of volitile N-nitrosamines in alcoholic beverages, *Food Cosmet. Toxicol.,* 17, 569, 1979.
44. **Druckrey, H., Preussmann, R., Ivankovic, S., and Schmahl, D.,** Organotrope carcinogene Wirkungen bei 65 verschiedenen N-nitroso-Verbindungen an BD-Ratten, *Z. Krebsforsch,* 69, 103, 1967.
45. **Magee, P. N. and Barnes, J. M.,** Carcinogenic nitroso compounds, *Adv. Cancer Res.,* 10, 163, 1967.
46. **Mohr, U.,** in Morphology of Experimental Respiratory Carcinogenesis, Netteshein, P. and Hanna, M. G., Jr., Eds., AEC Symposium Series 21, U.S. Atomic Energy Commission, USAEC Division of Technical Information Extension, Oak Ridge, Tenn., 1970, 353.
47. **Hoffmann, D., Schmeltz, I., Hecht, S. S., and Wynder, E. L.,** in *Polycyclic Aromatic Hydrocarbons and Cancer,* Vol. 1, Gelboin, H. and Tsu, P. O. P., Eds., Academic Press, New York, 1978, 85.
48. **Lu, A. Y. H.,** Multiplicity of liver drug metabolizing enzymes, *Drug Metab. Rev.,* 10, 187, 1979.
49. **Brunnemann, K. D., Yu, L., and Hoffmann, D.,** Assessment of carcinogenic volatile N-nitrosamines in tobacco and in mainstream and sidestream smoke from cigarettes, *Cancer Res.,* 37. 3218, 1977.
50. **Havery, D. C., Fazio, T., and Howard, J. W.,** Trends in Levels of N-Nitrosopyrrolidine in Fried Bacon, IARC Sci. Publ., No. 19, International Agency for Research on Cancer, Lyon, 1978, 305.
51. **Hoffman, D., Adams, J. D., Brunnemann, K. D., and Hecht, S. S.,** Assessment of tobacco specific N-nitrosamines in tobacco products, *Cancer Res.,* 39, 2505, 1979.
52. **Greenblatt, M. and Lijinsky, W.,** Nitrosamine studies: neoplasms of liver and genital mesothelium in nitrosopyrrolidine treated MRC rats, *J. Natl. Cancer Inst.,* 48, 1687, 1972.
53. **Greenblatt, M. and Lijinsky, W. J.,** Failure to induce tumors in Swiss mice after concurrent administration of amino acids and sodium nitrite, *J. Natl. Cancer Inst.,* 48, 1389, 1972.
54. **Hoffmann, D., Ranieri, R., Hecht, S. S., Maronpot, R., and Wynder, E. L.,** Effects of N'-nitrosonornicotine and N'-nitrosoanabasine in rats, *J. Natl. Cancer Inst.,* 55, 977, 1975.
55. **Hecht, S. S., Chen, C. B., Ohmori, T., and Hoffmann, D.,** Comparative carcinogenicity in F-344 rats of the tobacco specific nitrosamines N"-nitrosonornicotine and 4-(methylnitrosoamino)-1-(3-pyridyl)-1-butanone, *Cancer Res.,* 40, 298, 1980.
56. **Hilfrich, J., Hecht, S. S., and Hoffmann, D.,** Effects of N'-nitrosonornicotine and N'-nitrosoanabasine in Syrian golden hamsters, *Cancer Lett.,* 2, 169, 1977.
57. **Hoffmann, D., Adams, J. D., Brunnemann, K. D., Rivenson, A., and Hecht, S. S.,** Tobacco specific N-nitrosamines: occurrence and bioassays, in N-Nitroso Compounds: Occurrence and Biological Effects, Sci. Publ. 41, Bartsch, H., O'Neill, I. K., Castegnaro, M., and Okada, M., Eds., International Agency for Research on Cancer, Lyon, France, 1982, 309.
58. **Lijinsky, W. and Taylor, H. W.,** The effect of substituents on the carcinogenicity of N-nitrosopyrrolidine in Sprague-Dawley rats, *Cancer Res.,* 36, 1988, 1976.
59. **Lijinsky, W., Taylor, H. W., and Keefer, I. K.,** Reduction of rat liver carcinogenicity of 4-nitrosomorpholine by α-deuterium substitution, *J. Natl. Cancer Inst.,* 57, 1311, 1976.
60. **Baldwin, J. E., Branz, S. E., Gomez, R. F., Kraft, P., Sinskey, A. J., and Tannenbaum, S. R.,** Chemical activation of nitrosamines into mutagenic agents, *Tetrahedron Lett.,* 333, 1976.
61. **Chen, C. B., Hecht, S. S., and Hoffmann, D.,** Metabolic α-hydroxylation of the tobacco-specific carcinogen N'-nitrosonornicotine, *Cancer Res.,* 38, 3639, 1978.
62. **Hecht, S. S., Chen, C. B., and Hoffmann, D.,** Tobacco specific nitrosamines: occurrence, formation, carcinogenicity and metabolism, *Accounts Chem. Res.,* 12, 92, 1979.
63. **McCoy, G. D., Chen, C. B., and Hecht, S. S.,** Influence of modifiers of MFO activity on the *in vitro* metabolism of cyclic nitrosamines, in *Microsomes, Drug Oxidations and Chemical Carcinogenesis,* Vol. 2, Coon, M. J., Conney, A. H., Estabrook, R. W., Gelboin, H. V., Gillette, S. R., and O'Brien, P. J., Eds., Academic Press, New York, 1980, 1189.
64. **McCoy, G. D., Chen, C. B., and Hecht, S. S.,** Influence of mixed function oxidase inducers on the *in vitro* metabolism of N'-nitrosonornicotine by rat and hamster liver microsomes, *Drug Metab. Dispos.,* 9, 168, 1981.
65. **Harris, C. C., Autrup, H. H., Stoner, G. D., Trump, B. F., Hillman, E., Schafer, P. W., and Jeffery, A. M.,** Metabolism of benzo(a)pyrene, N-nitrosodiumethylamine, and N-nitrosopyrrolidine and identification of major carcinogen-DNA adducts formed in cultured human esophagus, *Cancer Res.,* 39, 4401, 1979.

66. **Autrup, H. and Stoner, G.**, Metabolism of N-nitrosamines by cultured human and rat esophagus, *Cancer Res.*, 42, 1307, 1982.
67. **Hecht, S. S., Reiss, B., Lin, D., and Williams, G. M.**, Metabolism of N'-nitrosonornicotine by cultured rat esophagus, *Carcinogenesis*, 3, 453, 1982.
68. **Lijinsky, W., Keefer, L., Loo, J., and Ross, A. E.**, Studies of alkylation of nucleic acids in rats by cyclic nitrosamines, *Cancer Res.*, 33, 1634, 1973.
69. **Scanlan, R. A., Farrelly, J. G., Hecker, L. I., and Lijinsky, W.**, Lack of metabolism of 2,6-dimethyldinitrosopiperazine by microsomes and post microsomal supernatant prepared from the rat esophagus and non-glandular stomach, *Cancer Lett.*, 10, 293, 1980.
70. **Protzel, M., Giardina, A. C., and Albano, U. H.**, The effect of liver imbalance on the development of oral tumors in mice following the applications of benzpyrene or tobacco tar, *Oral. Surg. Oral Med. Oral Pathol.*, 18, 622, 1964.
71. **Capel, I. D., Jenner, M., Pinnock, M. H., and Williams, D. C.**, The effect of chronic alcohol intake upon the hepatic microsomal carcinogen activation system, *Oncology (Basel)*, 35, 168, 1978.
71. **Stevens, M. H.**, Synergistic effect of alcohol on epidermoid carcinogenesis in the larynx, *Otalaryngol. Head Neck Surg.*, 87, 751, 1979.
73. **Elzay, R. P.**, Local effect of alcohol in combination DMBA and hamster cheek pouch, *J. Dent. Res.*, 45, 1788, 1966.
74. **Elzay, R. P.**, Effect of alcohol and cigarette smoke as promoting agents in hamster pouch carcinogenesis, *J. Dent. Res.*, 48, 1200, 1969.
75. **Freedman, H. and Shklar, G.**, Alcohol and hamster buccal pouch carcinogenesis, *Oral. Surg. Oral. Med. Oral Pathol.*, 46, 794, 1978.
76. **Seitz, H. K., Garro, H. J., and Lieber, C. S.**, Effect of chronic ethanol ingestion on intestinal metabolism and mutagenicity of benzo(a)pyrene, *Biochem. Biophys. Res. Commun.*, 85, 1061, 1978.
77. **Gibel, V. W.**, Experimentelle Untersuchungen zur Synkarzinogenese beim Osophaguskarzinom, *Arch. Geschwulstforsch.*, 30, 181, 1969.
78. **Schmähl, D.**, Investigations on esophageal carcinogenicity by methylphenylnitrosamine and ethanol in rats, *Cancer Lett.*, 1, 215, 1976.
79. **Girciute, L., Castegnaro, M., and Bereziat, J.-C.**, Influence of ethyl alcohol on the carcinogenic activity of N-nitrosdi-n-propylamine, in N-Nitroso Compounds: Occurrence and Biological Effects, IARC Sci. Publ. 41, Bartsch, H., O'Neill, I. K., Castegnaro, M., and Okada, M., Eds., International Agency for Research on Cancer, Lyon, France, 1982, 643.
80. **McCoy, G. D., Hecht, S. S., Katayama, S., and Wynder, E. L.**, Differential effect of chronic ethanol consumption on the carcinogenicity of N-nitrosopyrrolidine and N'-nitrosonornicotine in male Syrian golden hamsters, *Cancer Res.*, 41, 2849, 1981.
81. **Schrauzer, G. N., McGinness, J. E., Ishmael, D., and Bell, L. J.**, Alcoholism and cancer: effects of long-term exposure to alcohol on spontaneous mammary adenocarcinoma and prolactin levels in C3H/St mice, *J. Stud. Alchol*, 40, 240, 1979.
82. **Tuyns, A. J.**, Oesophagel cancer in non-smoking drinkers and non-drinking smokers, *Int. J. Cancer*, 32, 443, 1983.
83. **Wynder, E. L., Bross, I. J., and Feldman, R. M.**, A study of the etiological factors and cancer of the mouth, *Cancer*, 10, 1300, 1957.
84. **Wynder, E. L. and Fryer, J. H.**, Etiologic considerations of Plummer-Vinson (Patterson-Kelly) syndrome, *Ann. Intern. Med.*, 49, 1106, 1958.
85. **Larsson, L. G., Sandstrom, A., and Westling, P.**, Relationship of Plummer-Vinson disease to cancer of the upper alimentary tract in Sweden, *Cancer Res.*, 35, 3308, 1975.
86. **Cook-Mozaffari, P.**, The epidemiology of cancer of the esophagus, *Nutr. Cancer*, 1, 51, 1982.
87. **Hommozdiari, H., Day, N. E., Aramesh, B., and Mahboubi, E.**, Dietary factors and esophageal cancer in the Caspian Littoral of Iran, *Cancer Res.*, 35, 3493, 1975.
88. **Mahboubi, E. O. and Aramesh, B.**, Epidemiology of esophageal cancer in Iran, with special reference to nutritional and cultural aspects, *Prev. Med.*, 9, 613, 1980.
89. **Li, M.-X., Li, P., and Li, B.-R.**, Recent progress in research on esophageal cancer in China, *Adv. Cancer Res.*, 33, 173, 1980.
90. **Yang, C. S.**, Nitrosamines and other etiological factors in the esophageal cancer in northern China, in *Nitrosamines and Human Cancer*, Banbury Report No. 12, Magee, P. N., Ed., Cold Spring Harbor Laboratory, Cold Spring Harbor, N.Y., 1982, 487.
91. **Mellow, M. H., Layne, E. A., Lipman, T. O., Kaushik, M., Hostetler, C., and Smith, J. C.**, Plasma zinc and vitamin A in human squamous carcinoma of the esophagus, *Cancer*, 51, 1615, 1983.
92. **Graham, S.**, Results of case-control studies of diet and cancer in Buffalo, New York, *Cancer Res.*, 43, 24095, 1983.
93. **Meade, R. D., Yamashiro, S., Harada, T., and Basur, P. K.**, Influence of vitamin A on the laryngeal response of hamsters exposed to tobacco smoke, *Prog. Exp. Tumor Res.*, 24, 320, 1979.

94. **Leo, M. A. and Lieber, C. S.**, Interaction of ethanol with vitamin A, *Alcoholism: Clinical and Experimental Res.*, 7, 15, 1983.
95. **Leevy, C. M., Baker, H., tenHove, W., Frank, O., and Cherrick, G. R.**, B-complex vitamins in liver disease of the alcoholic, *Am. J. Clin. Nutr.*, 16, 339, 1965.
96. **DeLint, J.**, The prevention of alcoholism, *Prev. Med.*, 3, 24, 1974.
97. **Vitale, J. J. and Coffey, J.**, Alcoholism and vitamin metabolism, in *The Biology of Alcoholism*, Vol. 1, Kissin, B. and Begleiter, H., Eds., Plenum Publishing, New York, 1971, 327.
98. **Wynder, E. L. and Klein, V. E.**, The possible role of riboflavin deficiency in epithelial neoplasia. I. Epithelial changes of mice in simple deficiency, *Cancer (Philadelphia)*, 18, 167, 1965.
99. **Chan, P. C. and Wynder, E. L.**, The possible role of riboflavin deficiency in epithelial neoplasia. II. Effect on skin tumor development, *Cancer (Philadelphia)*, 16, 1221, 1970.
100. **Chan, P. C., Okamoto, T., and Wynder, E. L.**, Possible role of riboflavin deficiency in epithelial neoplasia. III. Induction of aryl carbon hydroxylase, *J. Natl. Cancer Inst.*, 48, 1341, 1972.
101. **Gerson, S. J. and Meyer, J.**, Increased lactate dehydrogenase activity in buccal epithelium of zinc-deficient rats, *J. Nutr.*, 107, 724, 1977.
102. **Fong, L. Y. Y., Sivak, A., and Newberne, P. M.**, Zinc deficiency and methylbenzylnitrosamine-induced esophageal cancer in rats, *J. Natl. Cancer Inst.*, 61, 145, 1978.
103. **Lin, H. J., Chan, W. C., Fong, L. Y. Y., and Newbern, P. M.**, Zinc levels in serum, hair and tumors from patients with esophageal cancer, *Nutr. Rep. Int.*, 15, 635, 1977.
104. **Harris, C. C., Spron, M. B., Kaufman, D. G., Smith, J. M., Jackson, F. E., and Saffioti, U.**, Histogenesis of squamous metaplasia in the hamster tracheal epithelium caused by vitamin A deficiency or benzo(a)pyrene and ferric oxide, *J. Natl. Cancer Inst.*, 48, 743, 1972.
105. **Salley, J. and Bryson, W.**, Vitamin A deficiency in the hamster, *J. Dent. Res.*, 36, 935, 1957.
106. **Dontenwill, W., Chevalier, H. J., Harke, H. P., LaFrenz, U., Reckzeh, G., and Schneider, B.**, Investigations on the effect of chronic cigarette smoke inhalation of Syrian golden hamsters, *J. Natl. Cancer Inst.*, 51, 1781, 1973.
107. **Kobayashi, N., Hoffmann, D., and Wynder, E. L.**, A study of tobacco carcinogenesis. XII. Epithelial changes induced in the upper respiratory tract of Syrian golden hamsters by cigarette smoke, *J. Natl. Cancer Inst.*, 53, 1085, 1974.
108. **Sporn, M. B., Dunlop, N. M., Newton, D. L., and Smith, J. M.**, Prevention of chemical carcinogenesis by vitamin A and its synthetic analogs (retinoids), *Fed. Proc.*, 35, 1332, 1976.
109. **Peter, F. M., Ed.**, The Health Effects of Nitrate, Nitrite and N-Nitroso Compounds, National Academy Press, Washington, D.C., 1981.
110. **Garcia, H. and Lijinsky, W.**, Studies of the tumorigenic effect in feeding of nitrosamino acids and low doses of amines and nitrite to rats, *Z. Krebsforsch.*, 79, 141, 1973.
111. **Greenblatt, M. and Lijinsky, W.**, Failure to induce tumors in Swiss mice after concurrent administration of amino acids and sodium nitrite, *J. Natl. Cancer Inst.*, 48, 1389, 1972.
112. **Chu, C. and Magee, P. N.**, Metabolic fate of nitrosoproline in the rat, *Cancer Res.*, 41, 3653, 1981.
113. **Ohshima, H. and Bartsch, H.**, Quantitative estimation of endogenous nitrosation in humans by monitoring N-nitrosoproline excreted in the urine, *Cancer Res.*, 41, 3658, 1981.
114. **Hoffmann, D. and Brunneman, K. D.**, Endogenous formation of N-nitrosoproline in cigarette smokers, *Cancer Res.*, 43, 5570, 1983.

Chapter 7

ALCOHOL, NUTRITION, AND CANCER

Nancy Misslbeck and T. Colin Campbell

TABLE OF CONTENTS

I. INTRODUCTION

Alcoholic beverages are used extensively worldwide, particularly in industrialized countires. These beverages are often consumed in excessive amounts over extended periods and there is some indication that consumption is increasing.[1] As constant exposure to any single substance could be expected to influence health, investigators have postulated a role for ethanol in the progression of chronic disease. Certainly excess ethanol consumption is one of the major causal factors in the development of cirrhosis, the fifth leading cause of death worldwide.[2] Consumption of ethanol may also be related to the development of specific malignancies, notably cancer of the upper GI tract, the liver, and the large bowel. Approximately 3 to 8% of all cancers in the U.S. have been attributed to ethanol consumption, although tobacco use by alcoholics is an important and probably the primary factor in the etiology of several of these cancers.[3,4]

A number of excellent articles have reviewed the epidemiological evidence for the associations among alcohol and tobacco use and the development of cancer.[3—9] The evidence linking ethanol to cancer of the GI, the large bowel, and hepatocellular carcinoma will be briefly reviewed in the following paper. Unfortunately, few laboratory experiments support the epidemiological evidence. The few long- and short-term studies that have been conducted will also be reviewed below. In this review, the question of whether ethanol is itself a carcinogen or whether it is a cocarcinogen will be addressed.

II. EPIDEMIOLOGICAL STUDIES

A. Cancer of the Upper GI Tract

The most significant association between excessive ethanol consumption and the development of cancer is with cancer of the upper GI tract, specifically cancer of the esophagus and larynx. This association was reported by Lamu[10] in 1910 when he noted an unusually high incidence of esophageal cancer in absinthe drinkers. Although congeners in the absinthe may have led to the formation of tumors in those cases, other studies have demonstrated the strength of that association.

Epidemiological studies in which per capita alcohol sales were correlated geographically with cancer mortality rates have demonstrated dose-dependent associations among alcohol consumption, tobacco use, and cancer of the upper GI tract. In the U.S., Breslow and Enstrom[11] found a strong correlation between cancer of the upper GI tract and consumption of spirits. Analysis of the data from 7518 personal interviews of individuals with invasive cancer in the Third National Cancer Survey showed significant positive associations between ethanol ingestion and cancer of the upper GI tract (pharynx, gum and mouth, larynx, lip, and esophagus). Significant associations were also found with the colon, rectum, breast, and thyroid. Suggestive associations were found with cancer of the liver, stomach, and pancreas. Cirrhosis, as an index of excessive ethanol intake, was significantly associated with cancer of the pharynx, esophagus, liver, pancreas, and lung. In Japan, Kono and Ikeda[12] found an association between consumption of shochu, a concentrated alcoholic beverage, and cancer of the esophagus. Hinds et al.[13] interviewed 9920 individuals to determine ethnic group habits of alcohol consumption and tobacco use in Hawaii. In their study, use of liquor and wine was correlated with pharyngeal cancer. Tobacco use was also strongly correlated with this type of cancer.

In a case-control study of various aspects of lifestyle of 1304 patients with cancer, Wynder et al.[14] found that a large proportion of those with oral cancer were heavy drinkers. Rothman and Keller[15] interviewed 598 men with squamous cell carcinoma of the mouth or pharynx and 598 controls. They found a dose-response relationship between ethanol consumption and cancer (r = 0.33). Both ethanol and smoking showed strong individual effects in their study. Graham et al.[16] in a case-control study at the Roswell Park Memorial Institute attempted to quantify the relationship between ethanol intake, tobacco use, and oral cancer. The subjects, 584 male patients with oral cancer and 1222 control patients, were carefully interviewed for information of alcohol and tobacco use and dietary intake. Oral hygiene was also examined and scored by dentists. Results indicated that smokers who drank had a 2.66 times greater risk of developing oral cancer than those who did not drink. Those who drank and smoked were at 15 times greater risk of developing oral cancer than the control population. Other case-control studies have demonstrated similar effects.[17—19] In these studies, oral cancer developed earlier in those who smoke and drank in excessive amounts.

Tobacco smoking is an important cofactor in the development of oral cancer in the alcoholic. Since tobacco smoke is such a potent carcinogen, the increased risk from ethanol is difficult to assess. From the studies cited above, it is not clear whether ethanol is itself a carcinogen or whether it is modifying the action of the tobacco smoke by increasing its potency and altering its primary site of action from the lung to the upper GI tract.

An increased risk due to ethanol consumption was found independent of tobacco smoking by the Addiction Research Foundation in Ontario, Canada.[20] In this prospective cohort study, age standardized death rates of 1823 known male alcoholics, who had attended alcoholism clinics, representing 85,641 person years, were compared to the expected mortality of the general male population in Ontario. Ethanol consumption, which in this study was approximately nine times greater than in the general population, was correlated with excess deaths from head and neck cancer in addition to heart disease, accidents and suicides, and cirrhosis. The association between these causes of death and excess alcohol intake remained high when corrections were made for tobacco use. A 3.3-fold increase in esophageal cancer was found in smokers who drank compared to smokers who abstained. The researchers concluded that although ethanol intake did not enhance the rate of cancer development in general, it apparently predisposed certain sites to cancer.

Hakulinen et al.[21] reported similar results in Finland. In their prospective study, the files from the cancer registry were matched with those of the "Ethanol Misuser Registry" (205,000 males convicted for alcohol-related offenses). The observed death rates were compared to the expected death rates in the male Finnish population. Esophageal and lung cancer were correlated with ethanol intake and smoking ($p \leq 0.001$).

The morbidity and mortality of Danish brewery workers, who have access to approximately 2 ℓ of beer a day while on the job, has also been studied.[22] Among the 14,313 workers who were located, cancer morbidity was increased for the pharynx, esophagus, liver, and larynx. As in the Canadian Study, mortality was not increased in this group compared to the general Danish population. Other cohort studies have shown similar results.[23]

From the studies cited here, ethanol may be a carcinogen, but it is more likely acting as a cocarcinogen. Three mechanisms have been proposed by which ethanol could exert its effects on the upper GI tract.[5,7,24] First, as an organic solvent, ethanol could facilitate the absorption of carcinogens, for example those from tobacco smoke, into the esophageal mucosa. Second, ethanol is an irritant and may destroy the mucosal barrier allowing carcinogens easier access to the epithelial cells.[25] Third, ethanol inhibits salivation. Reduced washing of the epithelial surface may allow local accumulations of carcinogens to form.[5] These mechanisms have not been examined experimentally, and therefore it is not known whether they actually contribute to pathogenesis of esophageal cancer.

The epidemiological studies have tended to support the hypothesis that ethanol is a cocarcinogen in the upper GI tract. In examining the results of these studies, however, several factors should be considered in addition to the traditional problems of interpreting epidemiological studies. First, many of the results are dependent on correlations with intake of a particular alcoholic beverage. For example, the correlations in France[26] are between esophageal cancer and apple brandy whiskey (Calvados) intake, and in Japan, the observations are dependent on rice wine consumption.[12] These beverages could contain carcinogenic congeners. Nitrosamines and other polycyclic aromatic hydrocarbons have been found in alcoholic beverages,[27–29] and aflatoxin may be present in rice wine.[12] Second, assessment of the quantity of alcohol consumed is difficult.[30] Studies in which consumption is assessed by questionnaire may therefore be inaccurate.[11,31] Third, alcoholics have often been reported to have a poor nutritional status due either to low intake of nutrients or to imbalances caused by the toxicity of ethanol. This factor may confound the results of the epidemiology studies.[5] Thus, although ethanol may play a role in cancer development, the data are difficult to interpret due to the presence other etiological factors which may influence the development of specific cancers.

B. Hepatocellular Carcinoma

The role of ethanol in the development of hepatocellular carcinoma is difficult to assess. Worldwide, approximately 60 to 90% of hepatocellular carcinoma is found in cirrhotic livers, and cirrhosis has been shown experimentally to predispose the liver to the development of hepatocellular carcinoma.[32] Chronic ingestion of ethanol has been linked to the development of cirrhosis. An estimated 10 to 15% of alcoholics develop cirrhosis.[33,34] Consequently, investigators have postulated a role for ethanol in the etiology of liver cancer.

Cirrhosis is not just a single disease. The pathology of alcoholic or fatty nutritional cirrhosis is quite different from that most frequently found in association with hepatocellular carcinoma worldwide. The cirrhosis found in countries with a high incidence of hepatocellular carcinoma is macronodular, the form often found following exposure to hepatitis B virus or hepatic toxins.[35] Micronodular cirrhosis is the form found in countries with high incidence of alcoholism. These countries typically have low rates of hepatocellular carcinoma in the population. The micronodular form may evolve into the macronodular form with repeated insult to the liver followed by repair.[36] This suggestion is supported by several studies in which alcoholics dying from hepatocellular carcinoma are approximately 5 to 10 years older than those dying from cirrhosis (see below).

The population studies of Kono and Ikeda in Japan,[12] Hinds in Hawaii,[13] and Breslow and Enstrom[11] in the U.S. indicate no increased risk of hepatocellular carcinoma when cancer incidence is compared geographically with per capita consumption of ethanol. The very low rate of hepatocellular carcinoma in these studies may have obscured a significant association.[4,34] In the Third national Cancer Survey, Williams and Horm[31] found a suggestive dose-dependent association between alcohol consumption and liver cancer. Cirrhosis was highly associated with liver cancer in this study ($p \leq 0.001$).

In the retrospective cohort study of Schmidt and Popham[20] which was described in the previous section, no increased risk of hepatocellular carcinoma in Canadian alcoholics was observed. The incidence of hepatocellular carcinoma was lower than expected in this study, given the high level of cirrhosis. The suggestion was made that the population may have been too young for hepatocellular carcinoma to have developed. Jensen's study of Danish Brewery Workers showed a significant increase in hepatocellular carcinoma ($p \leq 0.05$) compared to the general population.[22] Deaths from hepatocellular carcinoma were recorded in 29 individuals whereas only 19 were expected. Hakulinen et al.[21] also found a significant association between hepatocellular carcinoma and ethanol misuse in the Finnish population ($p \leq 0.05$).

No case-control studies have been conducted to examine the association between ethanol consumption and hepatocellular carcinoma. Due to the low incidence of hepatocellular carcinoma in countries where ethanol consumption is a significant problem, it is difficult to study large numbers of individuals with hepatocellular carcinoma. However, several reports indicate a high level of alcoholic cirrhosis in those dying from hepatocellular carcinoma.

In studies at the Boston City Hospital, Purtillo and Gottlieb[37] studied the results of 14,000 autopsies. Cirrhosis was found in 11.5% of that population and 0.69% had hepatocellular carcinoma. Hepatocellular carcinoma arose in 5% of those with cirrhosis. Of those dying from liver cancer, 79% had cirrhosis and 35% of those had the fatty nutritional cirrhosis associated with chronic ethanol consumption. They found an increase in the number of deaths from liver cancer and an increase in the cases of cirrhosis between 1917 and 1968.[37] MacDonald[38] had noted a similar trend after examining data from 1947 to 1954 in the same hospital. More recently, Omata et al.[39] reported that in their clinic, 22 out of 50 persons dying from liver cancer had alcoholic cirrhosis. In addition, it was noted that, on the average, those alcoholics dying of liver cancer were approximately 9 years older than those dying from cirrhosis. They suggested that the long-term damage to the liver by ethanol could result in liver cancer in those who do not succumb to cirrhosis. In another postmortem study of digestive tract cancers, 86 cases of hepatocellular carcinoma were found within a 3-year period in France.[40] Of these, 58 cases showed concomitant cirrhosis. Alcoholic cirrhosis was evident in 61% of cases of hepatocellular carcinoma. This incidence of hepatocellular carcinoma was considered high and a suggestion was made that the incidence is increasing.

Retrospective studies have examined the incidence of hepatocellular carcinoma in those dying from cirrhosis. In the extensive study of Norredam,[41] autopsies performed on 7763 individuals revealed 309 cases of cirrhosis of which 45 or 8% showed signs of hepatocellular carcinoma. Approximately 70% of those with hepatocellular carcinoma were elderly men. Johnson et al.[42] examined 294 patients with cirrhosis. Of those, 24% showed evidence of hepatocellular carcinoma. As in other studies, the development of hepatocellular carcinoma in cirrhotic livers was found in older individuals.

Lee[43] observed the incidence of hepatocellular carcinoma in those dying from alcoholic cirrhosis in the London Hospital in the 50 years between 1914 to 1963. Of the 182 cases of death from cirrhosis, 84 (46%) were clearly related to excessive ethanol consumption. Autopsies performed on those individuals revealed 24 cases of hepatocellular carcinoma (30%).

To summarize this information, although alcohol consumption is strongly associated with the development of cirrhosis and 60 to 90% of hepatocellular carcinoma is found in cirrhotic livers, about 10 to 15% of alcoholics develop cirrhosis, and of those only approximately 10% or less develop hepatocellular carcinoma. Perhaps as the population ages, the incidence of hepatocellular carcinoma in older alcoholics may become a more significant problem.

The association between ethanol intake and hepatocellular carcinoma is dependent on the prior development of cirrhosis. Whether cirrhosis is itself a preneoplastic state, whether it otherwise predisposes the liver to neoplasia is unknown. The link between ethanol and the development of cirrhosis is also controversial.[44] The hepatotoxic effects of ethanol may cause the scarring which leads to cirrhosis. Concomitant malnutrition or changes in the immune status of the individual may accompany or cause the formation of cirrhosis. Lieber et al.[24] have suggested that hepatocellular carcinoma may even arise in the noncirrhotic liver of those consuming excess ethanol. In that case, the ethanol itself may be acting as a direct carcinogen, although that is unlikely. Other possibilities are that metabolic changes could enhance the potency of some carcinogens, or that ethanol could enhance postinitiation development of hepatocellular carcinoma. These possible mechanisms need to be examined in the laboratory setting and will be discussed further in the sections on short- and long-term experiments.

C. Cancer of the Large Bowel

Several studies have indicated an association between consumption of ethanol and cancer of the large bowel. The incidence of colon cancer was correlated geographically in the U.S. with the consumption of beer in a study by Breslow and Enstrom.[11] Beer drinking was statistically associated with trends in socioeconomic status, with the sex ratio, and with urban-rural differences in rectal cancer.[45] In a similar study relating per capita consumption of alcoholic beverages to the incidence of cancer in Japan, Kono and Ikeda[12] found that wine consumption was associated with cancer of the rectum. The level of wine intake was low in this study. An assessment of ethanol consumption, tobacco use, and socioeconomic indicators in 7518 individuals with invasive cancer in the Third National Cancer Survey showed a correlation between intake of wine, beer, and hard liquor with cancer of the large bowel.[31] In that study, the association between ethanol and cancer of the upper GI tract was far stronger than with the large bowel. A time trend analysis of changes in food and beverage consumption in Australia with cancer of the rectum showed a positive correlation between beer consumption patterns and cancer.[46] However, nondietary factors were not considered in that analysis.

Other studies have not been able to show this positive relationship. Jensen conducted a retrospective study of causes of death and mortality in a cohort of danish brewery workers.[22] These workers, who were employed 6 months or longer by a brewery between 1939 and 1963, had access to approximately 2 ℓ of free beer per day. Of the 14,313 workers, data was obtained on 1303 cases of cancer. No difference was found in the observed incidence of colorectal cancer compared to the expected incidence. In a prospective study of male alcoholics in Ontario that was described earlier,[20] again no excess mortality from colorectal cancer could be attributed to ethanol consumption. Similarly, a correlational analysis by Hinds et al.[13] of types of ethanol consumed by ethnic groups in Hawaii and the cancer incidence revealed no association between beer drinking and colorectal cancer.

The inconclusive evidence on the association between cancer of the large bowel and the consumption of ethanol in general and beer in specific has led investigators to speculate that carcinogens present in beer may be responsible for the positive associations in some studies. The association has been found to show a strong geographical distribution.

III. SHORT-TERM TESTS

In support of the epidemiological evidence, ethanol could not be shown to be a carcinogen in short-term mutagenicity tests.[49] It was not mutagenic in the Ames assay, it caused no increase in recessive sex-linked mutations in *Drosophila melanogaster*, and it had no effect on sister chromatid exchange.[47] However ethanol did induce chromosomal aberrations in root tips of *Vicia faba* and other plant tissues. Acetaldehyde, the primary metabolite of ethanol, doubles the rate of sister chromatid exchange and may be responsible for the mutagenic effects observed following ethanol consumption.[48] Other short-term tests showed no effect of ethanol on chromsomal aberrations in Chinese hamster bone marrow cells, but chronic consumption did result in a lower aberration frequency caused by the direct-acting carcinogen patulin.[49] Again, the suggestion is that ethanol in some manner modifies the carcinogenic process, but does not itself interact with the genetic material in the cell.

In further support of the modifying activity of ethanol, ethanol ingestion has been shown to enhance the production of mutagenic metabolites from several carcinogens. Seitz et al.[50] reported that microsomes isolated from the intestines of rats that had consumed ethanol at 35% of caloric intake for 3 to 4 weeks had three times the benzo(*a*)pyrene hydroxylase activity of their pair-fed controls. In the Ames assay, the microsomes from the intestines of rats consuming ethanol produced more revertants with benzo(*a*)pyrene than did microsomes from control rats. The investigators postulated that increases in cancer incidence in alcoholics could be due to enhanced metabolism of procarcinogens in the intestine. Increased activation of procarcinogens has also been reported in the liver[51] and in the lung[52] with benzo(*a*)pyrene. McCoy et al.[53] found that microsomes from the livers of hamsters fed chronically with ethanol (35% of caloric intake in a liquid diet) showed enhanced metabolism of *N*-nitrosopyrrolidine. Capel et al.[54] reported similar effects of ethanol intake (as a 19% v/v solution in place of the drinking water) on carcinogen activating enzymes from mouse livers. However, Obidoa and Okolo[55] found no enhancing effect of ethanol consumption of the metabolism of aflatoxin B_1 (AFBl) by liver slices, and binding of the carcinogen to DNA was reduced.

The enhanced mutagenicity of microsomes isolated from animals administered ethanol is probably caused by induction of the components of the mixed function oxidase (MFO) system by chronic ethanol consumption. However, compounds which induce the MFO can also enhance detoxification, thus leading to more rapid excretion of a carcinogen from the cell. A carcinogen must remain bound to the DNA long enough for cell replication to occur in order to "fix" the lesion in the genome.[56-57] Therefore, even though ethanol may enhance activation in vitro, it may also protect the cell from initiation.

Another short-term method of measuring carcinogenicity is to assess macromolecular adduct formation.[58] Adduct formation can be influence by many factors that alter drug metabolism and disposition.[59] Although the relationship between the formation of adducts and the development of tumors remains unclear, the number of adducts formed is generally correlated with tumor development.[59] Schwartz et al.[60] measured the effect of ethanol consumption on formation of DNA adducts to ^{14}[C] dimethylnitrosamine (DMN). Rats were offered 20% w/v ethanol in the drinking water for 2 weeks; 24 hr after termination of the treatment, rats were injected with the carcinogen and sacrificed 4 hr later. No difference in the levels of specific adducts were found.

In a similar study, Belinsky et al.[61] examined the effect of ethanol consumption on replication, aklylation, and repair of DNA in liver cells treated in vivo with DMN. Rats were fed ethanol (35% of total caloric intake) in a liquid diet for 21 days prior to carcinogen treatment. Repair and alkylation of DNA following DMN treatment was not different in hepatocytes of animals consuming ethanol compared to controls. In addition, DNA replication was not enhanced by ethanol.

Mice consuming a 1% ethanol solution were found to have increased binding of benzo(*a*)pyrene to DNA for 20 weeks.[54] Ingestion of a 10% solution (v/v) actually decreased binding.

In an unpublished experiment from our laboratory, the effect of ethanol on the formation of macromolecular adducts to AFB_1 was assessed.[62] Fisher-344 rats (100 g) purchased from Charles River, Inc., Wilmington, Mass., were gradually acclimated to an AIN-76 based liquid diet containing 35% of the total calories from ethanol, were pair-fed an isocaloric control diet, or were offered control diet *ad libitum*. Rats were fed the diet for 5 weeks. On the day before the injection was given, food was removed at 5:00 p.m. Exactly 2 hr before injection, all rats were given 20 mℓ of food, which they consumed, to insure they would be in the same digestional state. Rats were then injected i.p. with 1 mg ^{3}H-AFBl/kg body weight (387 dpm/ng AFBl) (Makor Chemicals, Ltd., Jerusalem, Israel) dissolved in dimethylformamide (0.5 mℓ/kg body weight). Pair-fed, *ad libitum,* and ethanol-fed rats (five per group) were killed by decapitation at 2 hr after the injection. Livers were removed and weighed. The left medium lobe was frozen on solid CO_2 for analysis of adduct formation.

Although ethanol has been reported to induce the MFO drug metabolizing system, and AFBl requires metabolic activation by that system in order to bind to the cellular macromolecules, no effect on adduct formation was observed in this study (Table 1). The forms of cytochrome P_{450} that were induced by ethanol ingestion may not have influenced the metabolism of AFBl since different forms of P_{450} have different capabilities to metabolize AFBl. It is also possible that persistence of specific adducts may have been more significant in determining the relationship between adduct formation and carcinogenesis.[63]

Table 1
ADDUCTS, GLUTATHIONE, AND TOTAL RADIOACTIVITY

	Ethanol	Pair-fed control	Ad *libitum* control
Total (ng AFB1/g liver)	823 ± 135	776 ± 78	877 ± 103
DNA (ng AFB1/mg DNA)	11 ± 2	11 ± 2	12 ± 1.5
RNA (ng AFB1/mg RNA)	29 ± 5	27 ± 4	30 ± 5
Protein (ng AFB1/ng prot.)	1.6 ± 0.2	1.4 ± 0.2	1.5 ± 0.1
Plasma (ng AFB1/mℓ)	316 ± 43	311 ± 21	376 ± 39

Note: DNA, RNA, and protein adducts, plasma, and liver total radioactivity were determined in rats 2 hr after they had been injected with ^{3}H-aflatoxin B$_1$ (1 mg/kg body weight) dissolved in dimethylformamide 0.5 mg/kg). Rats were consuming diets containing ethanol (36% of total calories), were pair-fed control diet in which dextrin-maltose had been substituted isocalorically for the ethanol, or offered control diet *ad libitum.*

IV. LONG-TERM ANIMAL STUDIES

The carcinogenic potential of ethanol has also been examined in chronic feeding studies in experimental animals. However, few of these studies have been conducted. Ketcham et al.[64] offered ethanol in a 20% v/v solution in place of the drinking water to mice for up to 15 months. Although the mice lost weight initially on this regime, no change in longevity was observed and no increased incidence of tumors at any site was found. In the same study, ethanol intake had no effect on the growth or metastasis of transplanted tumors (from C. Cloudman S-91 melanoma inoculum) within 42 days of inoculation. Ketcham et al.[64] found no evidence of cirrhosis or fatty infiltration in the livers of their mice.

In another study in which ethanol was administered in place of the drinking water, Schmahl et al.[65] administered either brandy (38% of caloric intake) or a 25% ethanol solution to rats until their natural death and found no increased incidence of esophageal cancer. At this level of intake, ethanol itself did not appear to be carcinogenic. Studies to examine cocarcinogenic effects of ethanol also were negative. In the experiment cited above ethanol intake did not affect the mean induction period of esophageal tumors initiated by methylphenylnitrosamine. The investigators attributed the differences between their animal experiments and the human epidemiological evidence to the low intake of ethanol in the rat study. Again no sign of hepatotoxicity was found. They also postulated that either the congeners in alcoholic beverages or the concomitant presence of cigarette smoke could be necessary conditions for observing the cocarcinogenic effects of ethanol.

In another study by that same group, 270 rats were given diethylnitrosamine and ethanol concomitantly (5mℓ of a 25% solution per day) for their natural lifetimes.[66] The rats receiving ethanol actually had fewer liver tumors than the controls (35% of control rats developed liver tumors whereas tumors developed in only 5% of those consuming ethanol). However mortality was higher in rats treated with both diethylinitrosamine and ethanol.

In a preliminary study, Radike et al.[67] found that rats which recieved 5% ethanol in the drinking water while inhaling air containing 600 ppm vinyl chloride developed liver tumors earlier and more frequently than rats only inhaling vinyl chloride. Actual alcohol intake was not reported, but it was probably low. When ethanol is administered in the drinking water, intake is low compared to that of alcoholics.[68] In addition, dietary intake is not controlled when calories from ethanol are obtained in the drinking water. Reduced calorie intake might protect the animal against tumor growth. Development of a liquid diet in which ethanol provides up to 35% of total calorie intake[69] has allowed increased intake of ethanol in a nutritionally controlled diet.

Rats were fed an ethanol containing liquid diet for 14 months in a study by Mendenhall and Chedid.[70] From week 11 to 19, rats were dosed with 2 mg AFB1 per kilogram body weight. Death from toxicity during the dosing period was high in the group receiving both ethanol and AFB1 (11/26). Although hepatic peliosis was observed in 5/26 of the male rats consuming ethanol, no hepatocellular carcinomas were observed.

In another experiment female Sprague-Dawley rats consuming ethanol in a liquid diet were administered the carcinogen dimethylnitrosamine (DMN).[71] Rats were given the ethanol-containing diet for 3 weeks, and then given 1.5 mg DMN per day for 15 days while on solid control diet and *ad libitum* water. The control pellet feeding continued for 2 weeks before the animals were returned to the liquid diet. This cycle was followed 4 times for a total dose of 30 DMN. All rats were returned to solid food at 20 weeks for the duration of the experiment. By 90 weeks, all rats on the control diet had died whereas of those receiving ethanol, 4/17 (24%) were still alive. Although survival was increased by ethanol, (average survival time was 504 days vs. 565 days), no effect on the number or type of tumor was observed. Tumors were found predominatly in the liver and kidneys.

In these experiments ethanol was administered concurrently with the carcinogen thus eliminating the ability to determine which stage of carcinogenesis would be most affected by ethanol intake. In a study conducted in this laboratory, effects of ethanol on postinitiation development of presumptive preneoplastic lesions were examined.[81] Sprague-Dawley rats were pretreated with ten daily doses of the hepatocarcinogen, AFB1 (3.75 mg/kg body weight total dose), allowed to recover from the treatment, and then administered ethanol at 35% of kcals in a liquid diet for 12 weeks. Ethanol was administered in either low-fat (11% of caloric intake) or high-fat (35% of caloric intake) liquid diets the formulation of which was based on the AIN-76 recommendations.[73,74] A second set of rats was individually pair-fed to the rats receiving an ethanol diet in which dextrin-maltose had replaced the ethanol. A third group was given the nonethanol diet ad libitum. One additional group was administered diet containing ethanol for the first 8 weeks of postinitiation and then switched to the nonethanol diet for the final 4 weeks of the dietary treatment period. The purpose of this group was to test whether withdrawal from ethanol would affect development of hepatic lesions. After the rats had received the dietary treatment for 12 weeks, they were sacrificed. The formation of presumptive preneoplastic gamma glutamyltransferase (GGT)-positive foci was evaluated in each of the four lobes of the livers. Ethanol ingestion lowered hepatic vitamin A and glutathione levels, altered the hepatic phospholipid composition, and increased hepatic lipid levels when administered in combination with the high-fat diet. However, the number of foci found in the livers of rats consuming ethanol on either the low- or high-fat diet was not significantly higher compared to levels found in pair-fed or *ad libitum*-fed nonethanol controls (Table 2). In addition, withdrawal from ethanol did not alter the development of foci. Although ethanol did not influence the development of GGT-positive foci in this experiment, rats on the high-fat diet had significantly more foci than those on the low-fat diets ($p < 0.01$). It was concluded that the level of fat in the diet had a greater effect on the development of foci than did the chronic ingestion of ethanol.

Although ethanol did not appear to be a postinitiation modifier in this experiment, the species of animal which was used, the diet, the duration of the experiment, and the site of tumor development might have affected the results. First, the rat is not very susceptible to the toxic effects of ethanol. In this experiment, only those rats consuming ethanol in a high-fat diet developed fatty livers. Rats do not develop cirrhosis following ethanol consumption. Cirrhosis has been thought to predispose the liver to cancer.[32] Second, the AIN-76 formulation of the liquid diet used in this experiment provided in excess all nutrients necessary for the rat. A marginal diet might become nutritionally inadequate when coupled with ethanol intake. For example, animals consuming ethanol are more rapidly depleted of vitamin A and zinc stores than those not consuming ethanol.[75,76] Third, use of longer experimental periods and

Table 2
FOCI[a] PER CM³

	Low fat	High fat
Ethanol	23.4 ± 5.5	49.2 ± 12.2
Pair-fed	21.2 ± 3.9	40.3 ± 5.3
Ad libitum	12.7 ± 3.3	37.7 ± 6.8
Withdrawal		25.6 ± 4.2

[a] Statistical analysis by analysis of variance;
    the high-fat group had significantly more foci
    than the low-fat group; $p \leq 0.01$

the development of tumors as the endpoint might be necessary to show the influence of ethanol on hepatocarcinogenesis. Fourth, although ethanol has the most severe metabolic effects on the liver, other sites might be more susceptible to tumor development following ethanol intake.

Freedman and Klar[77] investiaged the effects of ethanol consumption on the development of papillomas in the cheek pouch of Syrian golden hamsters. The hamster consumes ethanol more readily than the rat and is more susceptible to the development of cirrhosis. In addition, cancer of the upper GI tract has been more consistently associated with ethanol consumption than hepatocellular carcinoma. Dimethylbenzanthracene (DMBA) was applied to the cheek pouch three times weekly as a 0.5% solution in mineral oil. Ethanol was administered as a 10% v/v solution in place of the drinking water. Those hamsters receiving both DMBA and ethanol developed larger and more invasive tumors approximately 2 weeks earlier than those receiving no ethanol. No differences in response between males and females were noted. However only three animals were sacrificed at each of three time points. Use of a liquid diet to administer the ethanol might have enhanced results. In an earlier study by Elzay,[78] bathing the cheek pouch with ethanol increased the number of DMBA-induced papillomas.

In another study with Syrian golden hamsters, McCoy et al.[79] examined the effects of chronic ethanol consumption in a liquid diet at 35% of total calorie intake on nasal cavity and tracheal tumors produced by *N*-nitrosonornicotine (NNN) and *N*-nitrosopyrrolidine (NPYR). Animals recieved the carcinogens in doses of 1 or 2 mmol 3 times weekly for 25 weeks while consuming ethanol or control liquid diets. Ethanol consumption enhanced the formation of nasal cavity and trachael tumors initiated by NPYR, particularly at the lower dose, but not by NNN. This difference was thought to be due to the difference in the site of— hydroxylation between the two compounds. Ethanol may have enhanced activation of a specific metabolic site.

These experiments support the epidemiological and short-term evidence that ethanol itself in not a carcinogen. Those few studies that examined consumption of ethanol at levels found in alcoholics showed that ethanol did not influence hepatocarcinogenesis but that it may have enhanced activation of specific carcinogens in the upper GI tract. Route of administration, the dosage of ethanol, the carcinogen, and the species of animals used all affected the outcome of these studies.

V. CONCLUSION

Evidence from epidemiological studies support the hypothesis that excessive consumption of ethanol can play an important role in the etiology of cancer of several sites. Although some of these studies indicate an independent effect of ethanol, in no study was a single risk factor isolated. Use of tobacco was primary factor in the majority of the cases. The individual roles of ethanol and tobacco have been difficult to distinguish because those who drink tend also to be smokers.[80] In addition, those who drink and smoke tended to do more of both than those who only smoke or only drink. The marked geographical distribution of some of the cases pointed to the confounding role of impurities or congeners. Other confounding variables included socioeconomic status and, most importantly, nutritional status. Estimations of consumption remained an additional problem in many of the studies. Actual consumption data was extremely difficult to obtain for use in case-control and prospective studies. National disappearance data may not have reflected actual consumption since it did not take into consideration all sources of alcoholic beverages. Since ethanol consumption was highly skewed, studies based on this type of data may not have accurately reflected the distribution of ethanol consumption within the population. More information on patterns of consumption should be obtained.

The cocarcinogenic activity of ethanol is confirmed by short-term tests. In none of these tests was ethanol itself carcinogenic. The effects of ethanol on cellular metabolism did result in increased production of mutagenic metabolites in some experiments. However, formation of macromolecular adducts was not altered by chronic ethanol consumption.

Ethanol was not a carcinogen in longer animal experiments. However, the suggestion from the short-term tests that ethanol could alter metabolism of carcinogens has not been tested. Ethanol does not appear to modify the postinitiation phase of hepatocarcinogenesis, although its effect on other sites has not been adequately examined. Unfortunately, no good animal model for the human alcoholic has been developed. The formulation of the liquid diet, which enabled investigators to administer up to 50% of total calories from ethanol to experimental animals, has increased the potential to study the role of ethanol in disease. However, this development is still fairly recent.

The major challenge which investigators face is to define the specific mechanism of ethanol-induced damage not only in carcinogenicity studies, but in all research on the chronic effects of excessive ethanol consumption. The toxic effects of ethanol itself, its effects on nutritional status, and the effect of poor nutritional status on the toxicity of ethanol all require extensive study before the mechanisms of the action of ethanol in the human can be understood. In addition, the possible mechanisms for the effect of ethanol on upper GI cancer, its properties as a solvent, its effects on the mucosal barrier, and its effects on salivation and the relation of these to the induction and development of cancer should be studied. Finally the role of ethanol in the development of cirrhosis and the subsequent effect of that disease on carcinogenesis also needs to be defined before the potential role for ethanol in carcinogenesis can be understood and more accurately estimated.

In conclusion, a relatively small number of total cancers, only 3 to 8%, can be attributed to excessive ethanol consumption.[1,4] Many of these are dependent on tobacco use. As tobacco use is responsible for such a large number of deaths every year, elimination of tobacco use would seem to be more important than defining the relatively minor role of ethanol.

REFERENCES

1. **Tuyns, A. J.,** Epidemiology of alcohol and cancer, *Cancer Res.,* 39, 2840, 1979.
2. **Tamburro, C. H. and Lee, H. -M.,** Primary hepatic cancer in alcoholics, *Clin. Gastroenterol.,* 10, 457, 1981.
3. **Tuyns, A. J.,** Cancer and alcoholic beverages, in *Fermented Food Beverages in Nutrition,* Academic Press, New York, 1979, 427.
4. **Rothman, K. J.,** The proportion of cancer attributable to alcohol consumption, *Prev. Med.,* 9, 174, 1980.
5. **Kissin, B. and Kaley, M. M.,** Alcohol and cancer, in *The Biology of Alcholism,* Vol. 3, Kissin, B. and Begleiter, H., Eds., Plenum Press, New York, 1974, 481.

6. **Flamant, R.,** Epidemiological research on the relationship between tobacco, alcohol and cancer, *Prog. Biochem. Pharmacol.,* 14, 36, 1978.

7. **McCoy, G. D. and Wynder, E. L.,** Etiological and preventive implications in alcohol carcinogenesis, *Cancer Res.,* 39, 2844, 1979.

8. **Schottenfeld, D.,** Alcohol as a co-factor in the etiology of cancer, *Cancer,* 43, 1962, 1979.

9. **Vitale, J. J., Broitman, S. A., and Gottlieb, L. S.,** Alcohol and Carcinogenesis, in *Nutrition and Cancer: Etiology and Treatment,* Newell, G. R., and Ellison, N. M., Eds., Raven Press, New York, 1981, 291.

10. **Lamu, L.,** Etude de statistique clinique de 134 cas de cancer de l'oesophage et du cardia, *Arch. Mal. Appar. Dig. Mal. Nutr.,* 4, 451, 1910.

11. **Breslow, N. E. and Enstrom, J. E.,** Geographic correlations between cancer mortality rates and alcohol-tobacco consumption in the United States, *J. Natl. Cancer Inst.,* 53, 631, 1974.

12. **Kono, S. and Ikeda, M.,** Correlation between cancer mortality and alcoholic beverages in Japan, *Br. J. Cancer,* 40, 449, 1979.

13. **Hinds, M. W., Kolonel, L. N., Lee, J., and Hirohata, T.,** Associations between cancer incidence and alcohol/cigarette consumption among five ethnic groups in Hawaii, *Br. J. Cancer,* 41, 929, 1980.

14. **Wynder, E. L., Bross, I. D. J., and Feldman, R. M.,** A study of etiological factors in cancer of the mouth, Cancer *(Philadelphia),* 10, 1300, 1957.

15. **Rothman, K. and Keller, A.,** The effect of joint exposure to alcohol and tobacco on risk of cancer of the mouth and pharynx, *J. Chronic Dis.,* 25, 711, 1972.

16. **Graham, S., Dayal, H., Rohrer, T., Swanson, M., Sultz, H., Shedd, D., and Fischman, S.,** Dentition, diet, tobacco, and alcohol in the epidemiology or oral cancer, *J. Natl. Cancer Inst.,* 59, 1611, 1977.

17. **Bross, I. D. J. and Coombs, J.,** Early onset of oral cancer among women who drink and smoke, *Oncology,* 33, 136, 1976.

18. **Burch, J. D., Howe, G. R., Miller, A. B., and Semenciw, R.,** Tobacco, alcohol, asbestos, and nickel in the etiology of cancer of the larynx: a case-control study, *J. Natl. Cancer Inst.,* 67, 1219, 1981.

19. **Keller, A. Z.,** Alcohol, tobacco, and age factors in the relative frequency of cancer among males with and without liver cirrhosis, *Am. J. Epidemiol.,* 106, 194, 1977.

20. **Schmidt, W. and Popham, R. E.,** The role of drinking and smoking in mortality from cancer and other causes in male alcoholics, *Cancer,* 47, 1031, 1981.

21. **Hakulinen, F., Lehtimaki, L., Lehtonen, M., and Teppo, L.,** Cancer morbidity among two male cohorts with increased alcohol consumption in Finland, *J. Natl. Cancer Inst.,* 52, 1711, 1974.

22. **Jensen, O. M.,** Cancer morbidity and causes of death among Danish brewery workers, *Int. J. Cancer,* 23, 454, 1979.

23. **Robinette, C. D., Hrubec, Z., and Fraumeni, J. F., Jr.,** Chronic alcoholism and subsequent mortality in World War II veterans, *Am. J. Epidemiol.,* 109, 687, 1979.

24. **Lieber, C. S., Seitz, H. K., Garro, A. J., and Warner, J. M.,** Alcohol-related diseases and carcinogenesis, *Cancer Res.,* 39, 2863, 1979.

25. **Fromm, D. and Robertson, R.,** Effects of alcohol on ion transport by isolated gastric and esophageal mucosa, *Gastroenterology,* 79, 220, 1976.

26. **Tuyns, A. J.,** Cancer of the oesophagus: further evidence of the relation to drinking habits in France, *Int. J. Cancer,* 5, 152, 1970.

27. **Masuda, Y., Mori, K., Hirohata, T., and Kuratsune, M.,** Carcinogenesis in the esophagus. III. Polycyclic aromatic mydrocarbons and phenols in whiskey, *Hannk,* 57, 549, 1966.

28. **Goff, E. U. and Fine, D. H.,** Analysis of volatile N-nitrosamines in alcoholic beverages, *Food Cosmet. Toxicol.,* 17, 569, 1979.

29. **Spiegelhalder, B., Eisenbrand, G., and Preussman, R.,** Contamination of beer with trace quantities of N-nitrosodimethylamine, *Food Cosmet. Toxicol.,* 17, 29, 1979.

30. **Mills, P. R., Shenkin, A., Anthony, R. S., McLelland, A. S., Main, A. N. H., MacSween, R. N. M., and Russell, R. I.,** Assessment of nutritional status and in vivo immune responses in alcoholic liver disease, *Am. J. Clin. Nutr.,* 38, 849, 1983.

31. **Williams, R. R. and Horm, J. W.,** Association of cancer sites with tobacco and alcohol consumption and socioeconomic status of patients: interview study from the Third National Cancer Survey, *J. Natl. Cancer Inst.,* 58, 525, 1977.

32. **Newberne, P. M., Harrington, D. H., and Wogan, G. N.,** Effects of cirrhosis and other liver insults on induction of liver tumors by aflatoxin in rats, *Lab. Invest.,* 15, 962, 1968.

33. **Lelbach, W. K.,** Cirrhosis in the alcoholic and its relation to the volume of alcohol abuse, *Ann. N.Y. Acad. Sci.,* 252, 85, 1975.

34. **Higginson, J.,** The geographical pathology of primary liver cancer, *Cancer Res.,* 23, 1624, 1963.

35. **Gall, E. A.,** Primary and metastatic carcinoma of the liver: relationship to hepatic cirrhosis, *Arch. Pathol.,* 70, 226, 1960.

36. **Farber, E.,** The pathology of experimental liver cell cancer, in *Liver Cell Cancer,* Cameron, H. M., Linsell, D. A., and Warwick, G. P., Eds., Elsevier, Amsterdam, 1976, 243.

37. **Purtilo, D. T. and Gottlieb, L. S.,** Cirrhosis and hepatoma occurring at Boston City Hospital (1917—1968), *Cancer,* 32, 458, 1973.
38. **MacDonald, R. A.,** Cirrhosis and primary carcinoma of the liver: changes in their occurrence at the Boston City Hospital, 1897—1954, *N. Engl. J. Med.,* 255, 1179, 1956.
39. **Omata, M., Ashcavai, M., Liew, C. -T., and Peters, R. L.,** Hepatocellularcarcinoma in the U.S.A.: etiologic considerations: localization of hepatitis B antigen, *Gastroenterology,* 76, 279, 1979.
40. **Faivre, J., Milan, C., Brignon, P., Legoux, J. L., Martin, F., and Klepping, C.,** Primary liver cancer in Cote d'Or (Burgundy): results of three years systematic registration in a well-defined French population, *Biomedicine,* 31, 150, 1979.
41. **Norredam, J.,** Primary carcinoma of the liver, *Acta Pathol. Microbiol. Scand.,* 87, 227, 1979.
42. **Johnson, P. J., Krasner, N., Portmann, B., Eddleston, A. L. W. F., and Williams, R.,** Hepatocellular carcinoma in Great Briton: influence of age, sex, HBsAg status, and aetiology of underlying cirrhosis, *Gut,* 19, 1022, 1978.
43. **Lee, F. I.,** Cirrhosis and hepatoma in alcoholics, *Gut,* 7, 77, 1966.
44. **Patek, A. J.,** Alcohol, malnutrition, and alcoholic cirrhosis, *Am. J. Clin. Nutr.,* 32, 1304, 1979.
45. **Enstrom, J. E.,** Colorectal cancer and beer drinking, *Br. J. Cancer,* 35, 674, 1977.
46. **McMichael, A. J.,** Alimentary tract cancer in Australia in relation to diet and alcohol, *Nutr. Cancer,* 1, 82, 1979.
47. **Obe, G. and Ristow, H. -J.,** Mutagenic, cacerogenic and teratogenic effects of alcohol, *Mutat. Res.,* 65, 229, 1979.
48. **Obe, G.,** Acetaldehyde not ethanol is mutagenic, *Prog. Mutat. Res.,* 2, 19, 1981.
49. **Korte, A., Slacik-Erben, R., and Obe, G.,** The influence of ethanol treatment on cytogenetic effects in bone marrow cells of Chinese hamsters by ethanol or patulin, *Toxicology,* 12, 53, 1979.
50. **Seitz, H. K., Garro, A. J., and Lieber, C. S.,** Effect of chronic ethanol ingestion on intestinal metabolism and mutagenicity of benzo(a)pyrene, *Biochem. Biophys. Res. Commun.,* 85, 1061, 1978.
51. **Seitz, H. K., Garro, A. J., and Lieber, C. S.,** Enhanced hepatic activation of procarcinogens after chronic ethanol consumption, *Gastroenterology,* 77, A40, 1979.
52. **Seitz, H. K., Garro, A. J., and Lieber, C. S.,** Enhanced pulmonary and intestinal activation of procarcinogens and mutagens after chronic ethanol consumption in the rat, *Eur. J. Clin. Invest.,* 11, 33, 1981.
53. **McCoy, G. D., Chen, C. B., Hecht, S. S., and McCoy, E. C.,** Enhanced metabolism and mutagenesis of nitrosopyrrolidine in liver fractions isolated from chronic ethanol-consuming hamsters, *Cancer Res.,* 39, 1979.
54. **Capel, I. D., Dorrel, H. M., Jenner, M., Pinnock, M. H., and Williams, D. C.,** The effect of prolonged ethanol intake on some carcinogen-activating enzymes in mice, *Biochem. Pharmacol.,* 28, 1139, 1979.
55. **Obidoa, O. and Okolo, T. C.,** Effect of ethanol administration on the metabolism of aflatoxin B_1, *Biochem. Med.,* 22, 145, 1979.
56. **Craddock, V. M.,** Induction of liver tumors in rats by a single treatment with nitroso compounds after partial hepatectomy, *Nature (London),* 245, 386, 1973.
57. **Craddock, V. M.,** Cell proliferation and experimental liver cancer, in *Liver Cell Cancer,* Cameron, H. M., Linsell, D. A., and Warwick, G. P., Eds., Elsevier, Amsterdam, 1976, 153.
58. **Wogan, G. N., Croy, R. G., Essigman, J. M., Groopman, J. D., Thilly, W. G., Shopek, T. R., and Liber, H. L.,** Mechanisms of action of aflatoxin B and sterigmatocystin: relationships of macromolecular binding to carcinogenicity and mutagenicity, in *Environmental Carcinogenesis,* Emmelot, P. and Kriek, E., Eds., Elsevier/North-Holland Biomedical Press, Amsterdam, 1979, 97.
59. **Lutz, W. K.,** *In vivo* covalent binding of organic chemicals to DNA as a quantitative indicator in the process of chemical carcinogenesis, *Mutat. Res.,* 66, 289, 1979.
60. **Schwarz, M., Wiesbeck, G., Hummel, J., and Kunz, W.,** Effect of ethanol on dimethylnitrosamine activation and DNA synthesis in rat liver, *Carcinogenesis,* 3, 1071, 1982.
61. **Belinsky, S. A., Bedell, M. A., and Swenberg, J. A.,** Effect of chronic ethanol diet on the replication, alkylation, and repair of DNA from hepatocytes and nonparenchymal cells following dimethynitrosamine administration, *Carcinogenesis,* 3, 1293, 1982.
62. **Misslbeck, N. G.,** The Effect of Chronic Ethanol Ingestion on the Development of Preneoplastic Lesions in the Liver, Ph.D. thesis, Cornell University, Ithaca, N.Y., 1983.
63. **Croy, R. G. and Wogan, G. N.,** Temporal patterns of covalent DNA adducts in rat liver after single and multiple doses of aflatoxin B_1, *Cancer Res.,* 41, 197, 1981.
64. **Ketcham, A. S., Wexler, H., and Mantel, N.,** Effects or alcohol in mouse neoplasia, *Cancer Res.,* 63, 667, 1963.
65. **Schmahl, D., Thomas, C., Sattler, W., and Scheld, G. F.,** Experimentelle Untersuchungen zur Syncarcinogenese. III. Mitteilung. Versuche zur Krebserzeugung bei Ratten bei gleichzeitiger Gabe von Diathynitrosamin und Tetrachlorkohlinstoff bzw. Athylalkohol: zugleich ein experimenteller beitrag zur frage der "Alkoholcirrhose", *Z. Krebsforsh.,* 66, 526, 1965.

66. **Habs, M. and Schmahl, D.**, Inhibition of the hepatocarcinogenic activity of deithylnitrosmine (DENA) by ethanol in rats, *Hepato-gastroenterology*, 28, 242, 1981.
67. **Radike, M. J., Stemmer, K. L., Brown, P. G., Larson, E., and Bingham, E.**, Effect of ethanol and vinyl chloride on the induction of liver tumors: preliminary report, *Environ. Health Perspect.*, 21, 153, 1977.
68. **Scheig, R.**, Effects of ethanol on the liver, *Am. J. Clin. Nutr.*, 23, 467, 1970.
69. **DeCarli, L. M. and Lieber, C. S.**, Fatty liver in the rat after prolonged intake of ethanol with a nutritionally adequate new liquid diet, *J. Nutr.*, 91, 331, 1967.
70. **Mendenhall, C. L. and Chedid, A.**, Peliosis Hepatis: its relationship to chronic alcoholism, aflatoxin B, and carcinogenesis in male Holtzman rats, *Dig. Dis. Sci.*, 25, 587, 1980.
71. **Teschke, R., Minzloff, M., Oldiges, H., and Franzel, H.**, Effect of chronic alcohol consumption on tumor incidence due to dimethylnitrosamine administration, *J. Cancer Res. Clin. Oncol.*, 106, 58, 1982.
72. **Gellert, J., Moreno, G., Haydn, M., Oldiges, H., Frenzel, H., Teschke, R., and Strohmeyer, G.**, Decreased hepatotoxicity of dimethylnitrosamine (DMN) following chronic alcohol consumption, *Adv. Exp. Med. Biol.*, 132, 237, 1980.
73. **American Institute of Nutrition**, Report of the Ad Hoc Committee on Standards for Nutritional Studies, *J. Nutr.*, 107, 1340, 1977.
74. **American Institue of Nutrition**, Report of the Ad Hoc Committee on Standards for Nutritional Studies, *J. Nutr.*, 110, 1726, 1980.
75. **Sato, M. and Lieber, C. S.**, Hepatic vitamin A depletion after chronic ethanol consumption in baboons and rats, *J. Nutr.*, 111, 2015, 1981.
76. **Ahmed, A. B. and Russell, R. M.**, The effect of ethanol feeding on zinc balance and tissue zinc levels in rats maintained on zinc-deficient diets, *J. Lab. Clin. Med.*, 100, 211, 1982.
77. **Freedman, A. and Sklar, G.**, Alcohol and hamster buccal pouch carcinogenesis, *Oral Surg.*, 46, 794, 1978.
78. **Elzay, R. P.**, Effect of alcohol and cigarette smoke as promoting agents in hamster pouch carcinogenesis, *J. Dent. Res.*, 48, 1200, 1969.
79. **McCoy, G. D., Hecht, S. S., Katayma, S., and Wynder, E. L.**, Differential effect of chronic ethanol consumption on the carcinogenicity of N-nitrosopyrrolidine and N-nitrosonornicotine in male Syrian golden hamsters, *Cancer Res.*, 41, 2849, 1981.
80. **Keller, A. Z. and Terris, M.**, The association of alcohol and tobacco with cancer of the mouth and pharnyx, *Am. J. Public Health*, 55, 1578, 1965.
81. **Misslbeck, N., Campbell, T. C., and Roe, D. A.**, Effect of ethanol consumption in combination with high and low fat diets on the post-initiation phase of hepato carcinogenesis, *J. Nutr.*, 114, 2311-2323, 1984.

Chapter 8

MUTAGENS AND CARCINOGENS IN FOOD

John H. Weisburger, William S. Barnes, and Richard Czerniak

TABLE OF CONTENTS

I. INTRODUCTION

New developments stemming from intensive research in the last 30 years on the mechanisms of cancer causation have revealed that the overall carcinogenic process can be resolved into a number of discrete steps.[1] The host-mediated biochemical activation of chemical carcinogens yields reactive electrophiles first called ultimate carcinogens by Miller and Miller.[2] Electrophiles react with the nucleophilic macromolecules in the cell, and the interaction with DNA is recognized to be an essential step. Certain direct-acting chemicals do not require this metabolic activation to react with DNA. Radiation can also be construed to interact with DNA in the same way as a direct-acting carcinogen. In addition, certain nucleic acid fragments from oncogenic viruses can be incorporated through appropriate mechanisms into the host DNA and lead to an altered DNA. Replication of the cells bearing such a modified DNA leads to the introduction of additional errors in DNA structure with consequent translocation of genes and their amplification.[3] It is thought that this sequence represents the molecular events connected with mutagenesis in prokaryotic and eukaryotic organisms. A mutational event does not necessarily lead to neoplastic transformation that may involve additional alterations in DNA structure and, perhaps, function.[4] In any case, these early events represent the "genotoxic pathway".[5] Once cells with such an abnormal DNA and a typical neoplastic gene product are present in sufficient numbers, they constitute the focus of neoplasia. Promotion may occur that enhances their growth by nongenotoxic pathways and mechanisms.[5-10] Whether or not an invasive neoplasm is obtained, therefore, depends on the overall process, with the early events and the later developments all playing essential roles.

For some kinds of cancer, such as stomach cancer, the genotoxic early events appear to play a major role, and promoting elements are not as important.[11,12] On the other hand, the complex constituents of tobacco smoke quantitatively resolve into relatively small amounts of a number of genotoxic carcinogens, such as polycyclic aromatic hydrocarbons and the tobacco-specific nitrosamines, but of greater importance is the fact that the promoting elements from the other constituents, such as phenols found in the acidic fraction, play a major role.[13] Because of these fundamental, distinct phases, cessation of exposure to tobacco smoke and, hence, removal of the forward pressure of promotion, leads to a progressively lower risk of lung cancer because the principal contributing event, promotion, is reversible and highly dose dependent.[14] Similar considerations apply to the important nutritionally linked cancers, such as cancer of the colon and cancer of the breast,[15,16] where the dietary fat level leads to promoting events through mechanisms discussed elsewhere in this book. Thus, lowering the fat intake should lower the promoting sequence and lead to lower disease risk,[17] as has already been documented in the smoking and cancer sequence.

Since estimates of the importance of diet in the development of many of the major cancers have ranged at 40 to 60% of cancer incidence,[18] examination of the responsible factors present in foodstuffs is a most promising line of inquiry in connection with cancer prevention. In this chapter, we will discuss the genotoxic elements in the human food chain in regard to cancers of the stomach, colon, and breast.

II. MUTAGENS AND CARCINOGENS IN FOOD

A. Polycyclic Aromatic Hydrocarbons

Polycyclic aromatic hydrocarbons, such as benzo(*a*)pyrene, have been observed in broiled or smoked fish or meats as well as in other foods.[19-22] There is no information as to whether these relatively small amounts of carcinogens, when ingested, would lead to human cancer. Larger amounts of such hydrocarbons, especially when administered in an edible oil vehicle, can induce forestomach cancer in mice or rats,[23] cancer of the intestines in sensitive strains of inbred hamsters,[24] and cancer of the mammary gland in certain strains of rats.[25] Polycyclic aromatic hydrocarbons that have been absorbed are excreted in the bile as such, or as metabolites. Small amounts of such hydrocarbons have been identified in the stool of rodents, but not of humans.[26] Thus far, intrarectal administration of these compounds in rats or mice has not induced colon cancer, but, as noted, oral or intrarectal administration to a sensitive strain of hamster produces colon cancer.[24] The reason for the strain or species difference is not known, but most likely depends on differences in biochemical activation and detoxification in the colon of rats and mice, vs. that of the inbred strain of hamster.

B. Polyhydroxylated Flavones

Many kinds of vegetables contain appreciable amounts of certain mutagenic, polyhydroxylated flavones, such as quercetin or kaempferol. The mechanism of their mutagenic activity in the *Salmonella* assay system is not exactly known, but it may be that in the presence of trace amounts of metal ions these chemicals yield hydrogen peroxide that would yield the noted mutagenic activity.[27,28] In mammalian cells, catalase serves to counteract hydrogen peroxide, and, thus, in such systems, a no-effect level should be observed as a function of dosage. Futhermore, in vivo, phenolic compounds are very effectively conjugated by Type II metabolic reactions to facilitate excretion.[29] When these detoxification systems are overloaded due to dose levels and length of exposure, then possible adverse effects may ensue. In most tests of quercetin, no evidence of carcinogenicity was noted.[30] In one series, liver cancer was reported,[31,32] but this may stem more from a promoting effect, as is true for other phenolic compounds, rather than from a carcinogenic effect.[33]

C. Aldehydes

Mutagenic activity has been observed in coffee. In a detailed inquiry as to the chemical nature of these mutagens, specific compounds identified were glyoxal, methylglyoxal, and diacetyl.[34] Whether these specific mutagens are carcinogens is unknown. Upon oral intake, however, these aldehydes could be expected to be rapidly detoxified by oxidation to the acid or reduction to the corresponding alcohol. Indeed, these chemicals have been noted to be part of the normal metabolism in living mammalian systems.[35]

D. Mutagens in Fried or Cooked Foods

Broiled and fried foods may be sources of genotoxic carcinogens associated with many of the nutritionally related cancers.[36,37] In view of the likely involvement in human cancer causation of the chemicals ensuing from such cooking processes, this aspect will be discussed in some detail.

Age-adjusted cancer mortality rates for the Western World show that in the last 40 years cancers of the colon and rectum, breast, and prostate have had only a slight increase.[38,39] This relative constancy in incidence of the nutritionally linked cancers reflects the fact that the mode of cooking and traditions in food intake and preparation have remained similar over this period in the Western world, but the situation appears to be different in Japan.[40-42] The traditional Japanese diets were quite different from Western diets in their mode of food preparation and their fat and salt content. However, in the last few decades, a slow progressive westernization has occurred with consequent alterations in the incidence of heart disease and colon and breast cancers.[43]

Carcinogens designated as genotoxic are mutagens in the Ames bacterial indicator system and induce DNA repair in the Williams liver cell test system.[44,45] Sugimura et al.[36] discovered that the surfaces of fried or broiled fish and meat developed potent mutagenic compounds for the Ames test. The structure of several of these mutagenic compounds has been determined (Figure 1). Structures of some of these compounds, 2-amino-3-methyl-imidazo-[4,5-f]quinoline (IQ) and the related compounds 2-amino-3,4-dimethylimidazo[4,5-f]quinoline (MeIQ) and

COMPOUND	STRUCTURE	MUTAGENICITY TA 98 REVERT/μG
2-AMINO-3-METHYL-IMIDAZO(4,5-F)QUINOLINE (IQ)	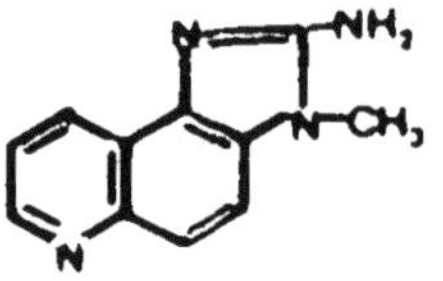	433,000
2-AMINO-3,8-DI-METHYL IMIDAZO (4,5F) QUINOLINE (MeIQ$_x$)	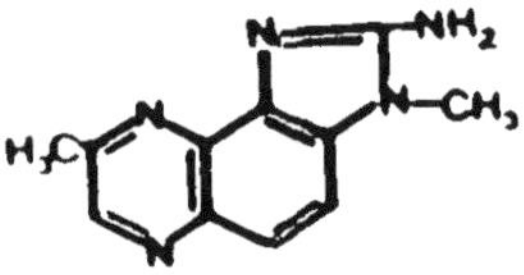	145,000
3-AMINO-1,4-DI-METHYL 5H-PYRIDO (4,3-B)INDOLE (TRP-P-1)	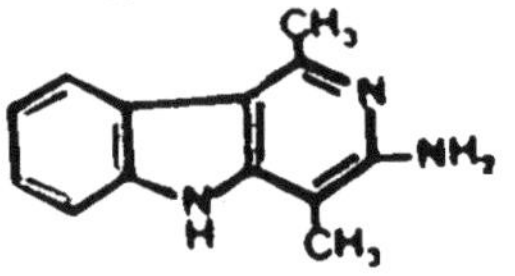	39,000
2-AMINO-α-CARBOLINE (AαC)	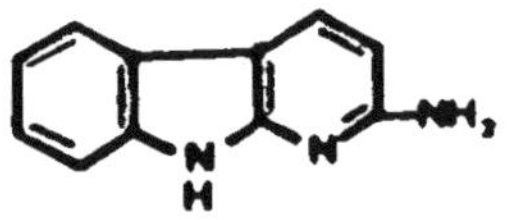	300
2-AMINO-3-METHYL-α-CARBOLINE (MEAαC)	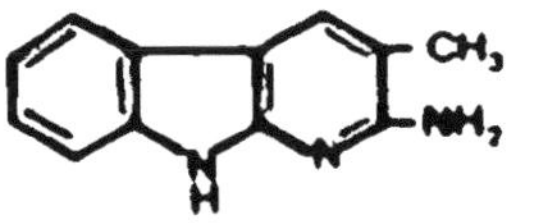	200
β-CARBOLINE (NORHARMAN)	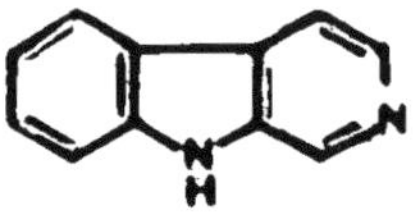	0
1-METHYL-β-CARBOLINE (HARMAN)	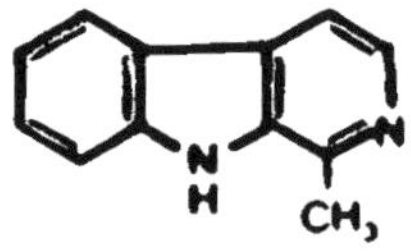	0

FIGURE 1. Structures of mutagens that have been identified in cooked beef and their mutagenicity for the Ames *Salmonella* typhimurium tester strain TA98 in the presence of rat liver S9.

2-amino-3,8-dimethylimidazo[4,5-f]quinoxaline (MeIQ$_x$), bear a structural resemblance to the potent colon, breast, and prostate carcinogen, 3,2'-dimethyl-4-aminobiphenyl (DMAB).[46] Williams found that these IQ-type compounds induce DNA repair in hepatocytes.[47] Others have found that these compounds are mutagenic in cultured mammalian cells[48—59] (Table 1).

Mutagen formation is itself a complex function of cooking time and temperature.[60,61] A lag period of approximately 5 min precedes the appearance of mutagenic activity. This is probably due to the temperature plateau reached at 100°C as the water content of the meat is reduced. Pariza found that meat can be cooked to the "medium well-done" stage without the formation of mutagens if cooking temperatures are kept low.[61] Frying is much more effective in producing mutagen formation than broiling. Microwave cooking does not result in any extractable mutagenic activity.[62]

A number of potent mutagens based on the imidazole-quinoline or imidazole quinoxaline ring systems have been found in cooked beef and in other cooked meats and fish. All of

Table 1
GENOTOXICITY OF FOOD MUTAGENS IN SHORT-TERM TESTS

Compound	Bacterial mutagenicity	DTᵃ/CHL mutagenicity[48,49]	6TGᵇ/CHO mutagenicity[50]	Chromosome aberations in vitro[50-52]	SCE in vitro[50,53—55]	Carcinogenicity	
						Mice[56,57]	Rats[58,59]
2-amino-3-methylimidazo [4,5-f]quinoline (IQ)	+	+	w	–	w	+	+
3-amino-1,4-dimethyl-5H-pyrido[4,3-b]indole (Trp-P-1)	+	+	ND	+	+	+	+
3-amino-1-methyl-5H-pyrido [4,3-b]indole (Trp-P-2)	+	+	+	+	+	+	+
2-amino-6-methyldipyrido [1,2-α:3,2'-d]imidazole (Glu-P-1)	+	+	ND	ND	+	+	+
2-amino-dipyrido[1,2-α:3,2'-d]imidazole (Glu-P-2)	+	+	ND	ND	w	+	+
2-amino-9H-pyrido[2,3-b]indole (2AαC)	+	+	ND	ND	+	+	ND
2-amino-3-methyl-9H-pyrido [2,3-b]indole (MeAαC)	+	ND	ND	ND	ND	+	ND

ᵃ Mutation to diphtheria toxin resistance in Chinese hamster lung cells.

ᵇ Mutation to 6-thioguanine resistance in Chinese hamster ovary cells. +, positive; w, weakly positive; -, negative; ND, no data.

Table 2
MUTAGENS IDENTIFIED IN SPECIFIC FOODS

Type of food	Mutagen identified[a]	Ref.
Broiled sardines, chicken, horsemackerel	Trp-P-1, Trp-P-2	64
Broiled herring, mackerel sardines	Trp-P-1, Trp-P-2	65
	IQ, MeIQ	
Broiled squid	Glu-P-2	66
Broiled sardines	IQ, MeIQ	65
Sun-dried sardines	Trp-P-1, Trp-P-2, IQ, MeIQ	67
Sun-dired cuttlefish	Glu-P-2	66
Broiled beef	Trp-P-1	66
Grilled beef, chicken, mushroom	AαC, MeAαC	63
Fried beef	MeIQ	68
Cereals, oils/fats, and all other food groups	Various PAHs	19
Coffee	Methylglyoxal, glyoxal, diacetyl	69
Various vegetables, fruits, berries	Flavanoids (quercetin, rutin, kaempferol)	70
Japanese pickles	Kaempferol, isorhamnetin	70

[a]   Mutagen identified by specific techniques described. It is probable that careful quantitative analysis of fried foods would demonstrate the presence, in variable amounts as a function of temperature, rate of heating, precursor, and modifier availability, of most of the mutagens belonging to this series of heterocyclic compounds.

these IQ-type chemicals have an exocyclic 2-amino group and are substituted by methyl groups on the imidazole portion or on the quinoline or quinoxaline ring system. In addition, another series of N-heterocyclic amines containing a carboline, or a closely related, ring system can be formed depending upon cooking conditions and the kind of meat or fish. These have been called Trp-type compounds after the first mutagen of this kind, 3-amino-1,4-dimethyl-5H-pyrido[4,3-b]indole (Trp-P-1), that was isolated from tryptophan pyrolysate. Trp-P-1 has been found in broiled beef, and 2-amino-carboline (AαC) and 2-amino-3-methylcarboline have been found in grilled beef.[63] Although not mutagenic themselves, harman and norharman enhance the activity of Trp-P-1 and other mutagenic carbolines and are also present in cooked beef.[34] A number of these compounds have been identified in other foods as well (Table 2).[64—70] Mutagen formation was also observed in other kinds of pan-fried fish.[71]

There is an interesting distinction in chemical properties between chemicals related to carbolines and the Trp-P series on the one hand, and the imidazoquinoline-quinoxaline group on the other hand. Treatment of these chemicals with nitrite at low pH invariably destroys the mutagenic activity of chemicals in the former group suggesting that the exocyclic amino group reacts in the conventional way, as would an aromatic amine. In contrast, the imidazoquinoline or quinoxaline series failed to react with nitrite under these conditions. We found when using pure IQ that the IQ is actually recovered unchanged. Reaction with nitrite permits a preliminary assessment of the presence of IQ-type compounds vs. other carbolines in a crude mixture of such chemicals stemming from fried or cooked foods.

1. Mode of Formation

Extensive Maillard reactions may yield mutagens and carcinogens. Model browning systems produce substances that are mutagenic in the Ames test with S9 added. When arabinose, 2-deoxyglucose, galactose, glucose, rhamnose, or xylose are refluxed with aqueous ammonium hydroxide, substances which revert Ames strain TA98 are produced. In another model browning experiment, the amino acids, rather than the sugars, were varied.[60,61,72] Equimolar amounts (0.01 *M*) of fructose or glucose and amino acid at a pH value of either 7 or 10, were autoclaved for 1 hr at 121°C and tested for mutagenicity.[73,74] These Maillard products also exhibited clastogenic activity in Chinese Hamster Ovary (CHO) cells.[75] Shi-

Table 3
MUTAGEN FORMATION IN A VARIETY OF FOODS[a]

Food	Sample	Cooking procedure	Cooking time(min)	Revertants/ sample	Revertants/ m^2
White bread	1 slice	Broiling	6	205	18,600
Pumpernickel bread	1 slice	Broiling	12	945	96,500
Biscuit	1 each	Baking	20	735	NC[b]
Pancake	1 each	Frying	4	2,500	153,000
Potato	1 small slice	Frying	30	200	329,000
Beef	1 patty	Frying	14	21,700	3,830,000

[a] Mutagen levels obtained after cooking food just beyond normal range of edibility. Basic fraction was tested on TA98 with S-9 as described in the text.

[b] No calculation due to the difficulty in measuring surface area of biscuits.

From Barnes, W. S., Spingarn, N. E., Garvie-Gould, C., Vuolo, L. L., Wang, Y.Y., and Weisburger, J. H., in *The Maillard Reaction in Foods and Nutrition*, ACS Symp. Series 215, Waller, G. R. and Feather, M. S., Eds., American Chemical Society, Washington, D. C., 1983, 485.

bamoto and associates[74] also found that browning reaction products of maltol and ammonia were positive in TA98 with S9. On the other hand, Aeschbacher et al.[76] found no bacterial mutagenicity in an arginine-glucose browning system.

The imidazole part of IQ compounds may arise from a common precursor. Matsushima heated equimolar amounts of ribose and imidazole or guanidine derivatives at 200°C, and those compounds, such as methylguanidine, creatine, and creatinine, that yielded mutagens all had a methylguanidine structure.[77] Jägerstad et al.[78] refluxed creatine, glucose, and amino acid in a model reaction system at 130°C that yielded mutagenic activity only with S9 activation, and in meat experiments found that the mutagenic activity, expressed as revertants per gram dry meatcrust, increased with increasing levels of creatinine. In subsequent experiments, MeIQ$_x$ was isolated from the model reflux system.

Foods with a high starch or sugar content may form genotoxic substances, but at a much lower level than meats or fish. Several common foods, in addition to beef, contained mutagens active for TA98 in the presence of S9.[60] Pariza et al.[61] found mutagenic activity in basic fractions of chicken broth, beef broth, rice cereal, bread crust, crackers, corn flakes, toast, and cookies. (Table 3).

2. Bioassays for Carcinogenicity

Carcinogenicity data are presently rather limited because the compounds concerned have been recently discovered. Tests of Trp-P-1 and Trp-P-2 (3-amino-1-methyl-5H-pyrido [4-3b] indole) in CDF1 mice yielded a high incidence of hepatocarcinomas, and the effect is sex specific, interestingly, with females being more sensitive. In male mice treated with Trp-P-2, there is also a slight, but statistically significant, excess of pulmonary adenocarcinomas.[34] Dietary administration of 2-amino-6-methyldipyrido [1,2α: 3′, 2′-d] imidazole (Glu-P-1) and 2-aminodipyrido [1,2-α: 3′, 2′-d] imidazole (Glu-P-2) to male and female mice induced heptocellular carcinomas, with females being again more sensitive. Me-α-carboline and α-carboline had the same effect, albeit the latter was somewhat less active with respect to induction of blood vessel tumors. Dietary administration of IQ to mice induced not only hepatocellular carcinomas, with females being more sensitive, but, interestingly, also forestomach and lung tumors.[57]

By analogy with aromatic amine carcinogens and heterocyclic carbolines such as Glu-P-1 and Trp-P-2, it has been assumed that activation of IQ to a mutagen proceeds by oxidation at the exocyclic amine. However, IQ is also an analog of quinoline, a liver and skin carcinogen

Group	Dose (mg/kg)	Total mammary tumors[a]
Diet control	—	0/9
Vehicle control		2/27
IQ	80 (0.4/mmol/kg)[b]	14/32[d]
4-AB[c]	20 (0.1/mmol/kg)[b]	19/32[d]

[a] In the IQ group also, other tumors were: liver tumors, 6 rats; ear duct tumors, 11 rats; leukemia, 2 rats; pancreatic tumors, 2 rats; urinary-excretory organs, 1 rat.

[b] The weight gain of the treated animals was about 93% of that of the vehicle controls, indicating that the dosage of IQ or of 4-AB being used was appropriate.

[c] The mammary carcinogen 4-aminobiphenyl (4-AB) was used as positive control.

[d] Significantly different from vehicle control group (P<0.05) by Fischer's probability test. See Reference 79.

in mice. In a recent experiment, IQ, quinoline, and DMAB were applied topically to shaved skin of Sencar mice, and after 20 weeks, 14 of 20 mice in the quinoline group had 25 tumors, but only 1 of 30 animals in the IQ group and 5 of 30 in the DMAB group were tumor bearing.[79] Hence, IQ appears to behave more as an aromatic amine than a quinoline, which would support the hypothesis that an *N*-hydroxy metabolite is the proximate genotoxic agent.[80]

After s.c. injections of Trp-P-1 into Syrian golden hamsters or Fischer rats, sarcomas were observed in 3 of 8 surviving hamsters and 5 of 20 surviving rats, but no tumors were present in animals injected with Trp-P-2 or in controls.[34] Skin painting experiments with Trp-P-1 and Trp-P-2 in female ICR mice were also negative. The induction of enzyme-altered foci in rat liver has also been studied in male Sprague-Dawley rats and in Fischer 344 males. Trp-P-1, Glu-P-1, and Glu-P-2 were positive (*p*<0.001) while Trp-P-2 was negative.[48,81]

Thus, Glu-P-1, Glu-P-2, AαC, and MeAαC have induced liver cancer in mice, with females being more sensitive. Mice serve as satisfactory test systems mainly because they require a smaller amount of test chemical for a given assay than do rats. However, many mouse strains are prone to develop liver neoplasms. The advantage of using rats is that they seem to mimic organotropism applicable to humans somewhat more faithfully. Thus, it is interesting that in rats Glu-P-1 not only induced liver cancer but also ear duct, brain, and colon neoplasms.[34] Trp-P-1 and Trp-P-2 fed to F344 rats induced hepatocellular carcinomas and intestinal cancers.

IQ has been tested in female Sprague-Dawley rats.[79] Within the relatively short period of 1 year, IQ induced a high incidence of cancer in the mammary gland and ear duct, and to a lesser extent, the liver, pancreas, and bladder (Table 4). A bioassay in male and female Fischer strain rats led to the induction of cancer in the intestinal tract, ear duct, liver, and pancreas.[59] Thus, IQ, a typical mutagen arising during cooking, can induce the kind of cancer (intestines, breast, and pancreas), so frequently seen associated with nutrition in the Western World.

3. Metabolism and Mode of Action

Metabolism of the mutagens found in fried meat is beginning to be studied by in vitro and animal experiments. Metabolic activation seems to be essential for this class of compounds as is true for carcinogenic arylamines. The cytochrome P-450 and especially the isozyme P-448 enzyme systems are effective in converting these chemicals to mutagens with the intermediary and hydroxy compounds being the likely proximate carcinogens. The further conversion of these compounds (as is true for the corresponding arylamines) can involve the formation of a sulfate ester and activation by an acetyl Co-A-dependent mechanism, presumably further requiring an NO-acetyl transfer.[82] In addition, the N-hydroxy Trp-P-2 appears to be activated by prolyl-t-RNA synthetase and seryl-t-RNA, but it seems that a similar compound Glu-P-1, was not so activated.[83]

Metabolism in mice or rats occurs after absorption from the intestinal tract or a parenteral injection site. Evidence suggests major metabolism in the liver. The parent compound and its metabolites are secreted into the bile and thence the intestinal tract. Some fraction of the metabolites are subject to enterobacterial metabolism, and the products reabsorbed as part of the enterohepatic cycle. Eventually, metabolites appear in urine and feces. Animals with biliary fistula exhibit no metabolites in stools, which indicates that absorption is complete.[84] Both urinary and biliary metabolites of Trp-P-1 and Glu-P-1 are complex and exist as free compounds and possibly as glucuronide conjugates.[85,86]

The reactive forms of Trp-P-2 and Glu-P-1 form adducts at the C-8 position of guanine through the exocyclic amino group.[87]

4. Modifying Factors in Mutagen Formation and Activity

Because IQ and similar mutagens are extensively consumed by humans in their diet, the mechanism of formation of these mutagens and any possible means of inhibiting such processes are under study. Beef suet added to lean ground beef gave increased mutagenic activity.[88] Lower concentrations (5 to 10% of initial wet weight) of corn oil enhanced mutagen formation more than lard, but the higher concentrations (20%) of both lipids approximately doubled the amount of mutagens (Figure 2). Different fats and oils have a similar augmenting effect, irrespective of the saturation level of the fat. This fat effect is not necessarily observed when different meats with distinct intrinsic fat contents are fried. Increases in temperature at the interface between pan and meat may contribute to these observed differences. Because lipids are responsible for the formation of flavor compounds and N-heterocyclics such as pyrazines and pyridines under real or simulated cooking conditions, they may also contribute precursors for mutagen formation.

We noted that glycerol also increased the formation of mutagens, a fact that may account for the effect of diverse fats since glycerol stems from the decomposition of fat. This, in turn, implies that Maillard-type reactions may be involved in the formation of mutagens, as will be discussed below.

Creatinine, pyrazines, and pyridines have been hypothesized to be precursors in mutagen formation. Addition of glycine and creatinine to ground beef prior to cooking enhances mutagen formation by approximately 50%. Supplementation with glycine, creatinine, and glucose or glycerol doubles the mutagenic activity. Mutagen formation is catalyzed by Fe^{3+} or Fe^{2+} that may be released through denaturation of heme proteins. Addition of 10 ppm Fe^{3+} and Fe^{2+} doubled mutagenic activity, but higher concentrations of Fe^{2+} were less effective. Concentrations of 10 to 60 ppm Fe^{3+} had similar enhancing effects.[89] The addition of ethylenediamine tetracetic acid (EDTA), a heavy metal chelator, when added to meat at levels of 1% (wt/wt) decreases mutagen formation during cooking to 50% of control values. Taylor has also noted an enhancing effect by iron.[90]

Additives such as soy protein, butylated hydroxyanisole (BHA), and chlorogenic acid decrease the amount of mutagenic activity formed during cooking (Figure 3). Other sub-

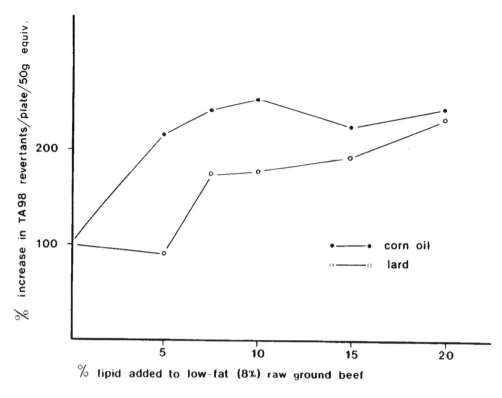

FIGURE 2.   Inhibition of mutagen formation by soy protein concentrates, butylated hydroxyanisole (BHA), and chlorogenic acid when added to ground beef prior to cooking.

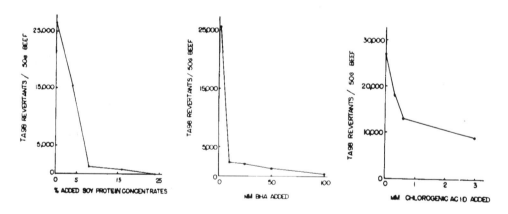

FIGURE 3.   Increase in mutagen formation with the addition of unsaturated (corn oil) or saturated (lard) lipids to ground beef prior to cooking.

stances inhibit the in vitro mutagenicity of pyrolysis products, and some of these are hemin, chlorophyllin, chlorophyl, and aqueous extracts from vegetables such as cabbage, radish, turnip, and ginger.[91-93] Fatty acids, especially oleic acid, also inhibit mutagenicity when added to samples of basic beef extract for Ames testing.[91] The role of oleic acid in inhibition is now thought to involve competition with Trp-P-2 for oxidation by microsomal cytochrome P-448. Pariza has also observed an inhibition for the development of mutagenicity in beef.[61]

5. Comment

Sugimura's pioneering finding of mutagens in fried foods has provided a powerful stimulus in cancer research and in nutritional studies. Whereas, the public has often been concerned and, indeed, alerted to synthetic contaminants in the food chain, it may well be that the studies currently in course in many laboratories will provide a new assessment of the mode of cooking as a possible source of hazardous compounds. It is premature in the light of current knowledge to draw definite conclusions but that these mutagens and carcinogens for several important target organs, not only liver but colon, breast, and others, are typical of aromatic amines known to be hazardous to man is quite striking. If any regulatory agencies in any country were made aware of a synthetic chemical or food additive with properties such as these compounds, arising from the traditional ways of cooking, it seems reasonably certain that such authorities would take action even with the data now at hand to limit human exposure or in fact ban such carcinogens. Thus, it is essential in the next few years to develop a sound data base as to the role of these chemicals formed during cooking and possibly leading to important human diseases.

It is most interesting to find not only that mutagens are present in many common cooked foods but that in most cases they are some of the most potent mutagens so far identified in the bacterial tester strains of Ames. It is important to realize that, for a number of reasons, it is difficult at this time to translate these results alone into an estimate of human risk. In regard to human risk, if indeed this kind of mutagen affects the colon and breast,[37] any effect would be enhanced by promoting factors.[94-99] Thus, the mechanisms of colon and breast cancer induction are in need of detailed research. Although genotoxic agents are likely to be important, promotion is also intimately involved. In fact, a great deal of effort at our Institute has been directed at this question.[100]

E. Mutagens in Pickled or Smoked Foods

1. Application to the Etiology of Gastric Cancer

Cancer of the glandular stomach occurs in high incidence in Japan, Iceland, mountainous interior regions of Central and Western Latin America, and some Eastern European countries. In contrast, Western Europe, many Anglo-Saxon countries (except Wales in the U.K), and the U.S. have a low incidence of gastric cancer. Over the last 50 years, the gastric cancer rate in the U.S. has steadily declined from 30 to 3.7/100,000 for men and 22 to 7.5/100,000 for women.[38,101-105]

Positive correlations with dietary risk factors in high-risk populations have been made for diets with a high consumption of dried, salted fish; pickled vegetables; smoked fish, and a low vegetable intake as well as for a reduced vitamin C intake, particularly on a seasonal basis.[12,43,101,106] Another correlation has been made between elevated levels of nitrate in foods and drinking water that occur due to high levels in the soil and water supply. The latter correlation holds only in the periodic absence of fresh vegetables as sources of vitamin C, antagonist of nitrite.[107,108] When fresh vegetables and fruits are more widely available, measurement of urine nitrate alone fails to correlate with risk. Crude salt or saltpeter used in the preservation of certain foods such as fish, may constitute another source of nitrate.[101,109-111] Japanese migrants to Hawaii whose diet contained a higher intake of uncooked vegetables such as celery, lettuce, tomatoes, and fresh fruit juices, had a lower risk of gastric cancer as compared to indigenous Japanese and first generation Japanese migrants to Hawaii.[40,43,112]

Alklylnitrosoureido compounds such as N-methyl-N'-nitro-N-nitrosoguanidine (MNNG) are the specific carcinogens with the characteristic of inducing glandular stomach cancer in animal models.[113] Sander found that such carcinogens were formed through the reaction of nitrite and suitably substituted secondary amines or amides.[114] The reaction of the nitrosation of alkylamides such as methylurea has been studied by Mirvish,[12] who also made the important discovery that ascorbic acid had an inhibitory effect on the nitrosation of methylurea and alkylamines.

1-Methyl-1,2,3,4-tetrahydro-β-carboline-3-carboxylic acid, isolated from soy sauce, when treated with nitrite under acidic conditions mimicking stomach acidity, led to direct-acting mutagenic activity.[115] This chemical is not the expected nitroso compound, which turned out to be not mutagenic, but is an as yet unknown product. Recently, tyramine has been isolated as another promutagen from soy sauce. When tyramine was nitrosated, 4-(2-aminoethyl)-6-diazo-2,4-cyclohexadienone, a direct-acting mutagen, was isolated, and studies on its carcinogenicity are underway. Interestingly, the higher the tyramine concentration of the soy sauce, the greater the mutagenicity upon nitrosation. At the Massachusetts Institute of Technology (MIT),[116] similar experiments were conducted with fava beans, often eaten in parts of Latin America, where gastric cancer occurs relatively frequently. In this instance, direct-acting mutagenic activity was also found. Yano discovered alkylating activity in several kinds of fish,[117] including mackerel, treated with nitrite. Glutathione may play a role in the release of reactive products in the glandular stomach.[118,119] A chemical found in the commercially cultivated mushroom, *Agaricus bisporus*, agaritine, β-*N*-[-L(+)-glutamyl]-4-hydroxymethylphenylhydrazine, yields 4-(hydroxymethyl)phenylhydrazine upon hydrolysis, and the corresponding diazonium compound induced glandular stomach cancer in Swiss mice.[120]

In our laboratory, we have tested certain types of fish, such as Sanma or Aji, frequently eaten in the Orient in high-risk regions for stomach cancer, as well as beans that are consumed in Latin America, and borscht which is consumed in Eastern Europe. For all of these products, direct-acting mutagenic activity was found. We noted that formation of the mutagen was blocked by the simultaneous presence of vitamin C. A mutagenic extract from the reaction of nitrite and Sanma induced adenocarcinomas of the glandular stomach in Wistar rats (Table 5).[107] The precursor found in fish is a rather polar compound, most likely with a carboxyl or hydroxy group, and of course, a nitrosatable amino group, to yield the reactive, direct-acting mutagen. These data are summarized in Table 6.[107,121]

All of these different lines of evidence lead to the conclusion that the human consumption of salted and pickled foods may result in gastric cancer through the action of an active agent (or agents) formed from nitrite and undefined substrates.[11,101,104] The fact that the formation of such carcinogens may be blocked by vitamin C or by vitamin E is most important. Thus, the formation of genotoxic gastric carcinogens may be inhibited by an intake of foods containing vitamin C with each meal all year long from childhood onward. The reduced intake of salted, pickled foods that most likely contains such gastric carcinogens may be another important element.[107] These factors may account for the sharp decline in the incidence of gastric cancer in the U.S., a decline also beginning to be apparent in other areas of the world where gastric cancer is still a major cause of mortality.[12,103-105] A reduction in the use of salt would have similar benefits and also reduce the risk of hypertension. In the MNNG-induced gastric cancer model, salt (sodium chloride) has a cocarcinogenic effect.[122,123] This observation may also mimic the human environment, for Joossens and Geboers[104] have documented a parallelism in international trends between gastric cancer and stroke. On the other hand, Kono et al.,[124] while finding a strong association between salt use and hypertension and stroke in a study of disease incidence in 46 prefectures and 12 regions within Japan, failed to find a correlation with gastric cancer. This result may be consistent with the concept of salt as a potentiator or cocarcinogen in human gastric cancer, rather than being the initial causative stimulus leading to the early events associated with gastric carcinogenesis, such as atrophic gastritis and intestinal metaplasia. This, in turn, suggests that the carcinogen is already preformed in pickled and smoked foods, with salt acting mainly as an enhancing element.

Under these conditions, the protection afforded by yellow-green vegetables and fruits as noted in migrant studies,[125] or the observations of gastric cancer areas in Japan,[43] may suggest

Table 5
SITES AND INCIDENCE OF TUMORS IN RATS GIVEN EXTRACTS OF FISH TREATED WITH NITRITE OR FISH ALONE

Site and type of tumor	Number of rats in experimental group I (fish extract alone)	Number of rats in experimental group II (fish extract + NaNO$_2$)
Effecitve number of rats	8[a]	12[b]
Forestomach		
Papilloma	0	0
Squamous cell carcinoma	0	2[c]
Glandular stomach		
Adenoma	0	2[d]
Adenocarcinoma	0	2[e]
Adenosquamous carcinoma	0	1[f]
Pancreas		
Adenoma	0	2
Adenocarcinoma	0	1
Small intestine		
Adenocarcinoma	0	1[f]

[a] Three rats had interstitial cell ademona of testis, 1 renal nephroblastoma, 1 rhabdomyosarcoma.
[b] Eight rats had at least one of the tumors tabulated here. In addition, the following miscellaneous tumors were noted: 1 rat had cortical adenoma of adrenal gland, 1 had follicular adenoma of thyroid gland, and 1 had pulmonary adenoma.
[c] One rat also had a pancreatic adenoma, and the other had an adenocarcinoma.
[d] One rat had only an adenoma, but the other also had an adenocarcinoma.
[e] One rat also had an adenocortical adenoma, and the other had a gastric adenoma and a thyroid adenoma.
[f] The only tumor in 1 rat.

From Scanlan, R. A. and Tannenbaum, S. R., Eds., *N-Nitroso Compounds*, ACS Symp. Series 174, American Chemical Society, Washington, D. C., 1981.

that the vitamin A or β-carotene content of such foods, in addition to the vitamin C and vitamin E content, deserves consideration.[126] Vitamin A plays a role in differentiation and other elements inhibiting the development of neoplasia.[127] Thus, the relevant mechanisms need exploration.

2. *Esophageal and Liver Cancer*

In areas of the world such as eastern Iran and central China, the high incidence of esophageal cancer does not appear to be related to the heavy use of tobacco and alcohol as it is in Western countries.[128,129] In Iran the poor quality of the diet with a low intake of fresh vegetables and fruits, the possible smoking of opium-tobacco, or the chewing of the resulting tar appear to be contributing factors.[129,130]

Consumption of traditional preserved foods such as salted fish and pickled vegetables among the southern Chinese people has been associated with their high incidence of nasopharyngeal and esophageal cancers.[39,101,131-135] The salted fish is commonly "pickled" in a crude salt solution of rock or sea salt containing nitrate. Significant levels of *N*-nitrosodimethylamine (NDMA) and other nitrosamines have been detected in salted fish. Rats fed salted fish daily produced urine that showed mutagenic activity in the Ames test.[136] Extracts from pickled vegetables that are commonly contaminated with fungus showed mutagenic, transforming, and promoting activity in different biological systems. From ethereal extracts of pickled vegetables, Lu et al.[137] have isolated a tetranitroso compound, Roussin's red

Table 6

**FOOD EXTRACTS DIRECTLY MUTAGENIC FOR THE AMES
SALMONELLA TYPHIMURIUM TESTER STRAINS AFTER
TREATMENT WITH NITRITE[107,121]**

Food extract	*S. typhimurium* strain TA	Dose (g/eg)[a]	Induced his + revertants[b]
Saltwater fish			
Sanma-hiraki (*Coloabis saira*)	1535	1	595
Aji-hiraki (*Trachurus japonica*)	1535	1	610
Marubushi-iwashi	1535	1	1,320
Canned fish			
Jack mackeral, dark muscle	1535	1	55
Herring	1535	1	63
Sardines	1535	1	72
Other food products			
Soy sauce	100	1	25,200
Fava beans	100	1	2,100
Fish sauce (Japanese)	100	1	2,300
Fish sauce (Philippine)	100	1	3,000
Fish sauce (Thai)	100	1	9,400

[a]   Dose is measured as the amount of food extract that is equivalent to the initial wet weight of food in grams.

[b]   Induced revertants = treated revertants − spontaneous revertants. All values shown are positive as measured by a doubling of revertants over spontaneous values.

methyl ester $[(NO)_2Fe(CH_3S)]_2$, that was mutagenic in *Salmonella* tester strain TA100 with a liver activation system.

In Iran, southern Soviet Union, and central China, the specific carcinogens are not yet known and may well be different for each area. In China, the compounds may be in the diet, or formed from dietary constituents, in which case vitamin C or vitamin E, or foods containing these micronutrients may be effective inhibitors.

III. APPLICATION TO WORLDWIDE CANCER PREVENTION

Nutrition and specific nutritional components, as well as dietary traditions, in various parts of the world may play an important role in the causation and development of a number of types of major human neoplasms. Nutrition may relate directly to the occurrence of 30 to 40% of cancers in men and 50 to 60% of cancers in women in the U.S. and other Western countries. The rapid, partial Westernization of nutritional customs in Japan has led to parallel alterations in the incidence of specific cancers.[43] On the other hand, the traditional salting and pickling procedures in Asia lead to different kinds of carcinogens, increasing the risk for cancers of the head and neck and liver.

As noted, the carcinogenic process involves a number of sequential steps, necessary for a clinically invasive cancer to occur. We have discussed herein mutagens and carcinogens as the agents possibly associated with the initiation of neoplastic changes for such cancers as colon, breast, prostate, and perhaps even pancreas. Such agents are found at the surface of fried or broiled foods such as meat or fish. While epidemiologic studies of meat-eating populations show an association with specific kinds of cancer, there are clearly certain inconsistencies. Indeed, the colon cancer incidence of vegetarian populations is consistently lower than that of meat-eaters; the incidence for breast cancer is only slightly lower and that of prostate cancer is similar. For colon cancer, the lower incidence can also be accounted for by a higher intake of fiber. On the other hand, at least in the Western world, many vegetarians adopt this lifestyle as adults. Exposure to genotoxic carcinogens early in life, especially during the phase of rapid growth, and proliferation of tissues such as breast or prostate during puberty, may suffice to increase the risk of cancer development later in life. For example, the single radiation exposure of the first atomic bomb dropped in Hiroshima showed that girls of the age group, 10 to 13, had a fourfold higher risk of breast cancer than did older women.

A fraction of adults on a Western diet have specific mutagens in their feces, whereas other nonmeat-eating groups exhibit none. One type of such mutagens appears to have a bacterial origin,[138-143] and its chemical structure is known.[144] Part of the fecal or urinary mutagenic activity may also be due to metabolites of mutagens from fried foods, an area requiring further research.

Gastric cancer appears to have distinct risk factors, namely, high consumption of pickled and salted fish, of beans, or residence in areas with geochemical or agricultural sources of nitrate intake, not balanced by the presence of vitamin C, vitamin E, or certain phenolic antioxidants and nitrite traps such as propyl gallate or tannins. The formation of the possible genotoxic carcinogen from a reaction with nitrite is inhibited by vitamin C, vitamin E, and certain antioxidants. This fact can be used as a preventive approach to deliberately decrease the risk for gastric cancer. Furthermore, cancer of the oral cavity and esophagus, highly prevalent in some parts of the world, appears to depend on the presence of certain nitrosamines, the amount of which can also be lowered by vitamins C or E.

Thus, it seems essential to conduct research on optimal levels of vitamins, minerals, antioxidants, and other micronutrients in the current diet in order to develop a broad base for chemoprevention. Also, if the work reviewed in this article can be extended to realistic actual human environmental situations, it would seem that the mode of food preparation, cooking by frying and broiling or by salting and pickling, may yield carcinogens for the major kinds of nutritionally linked cancers. Using approaches through which the formation of such carcinogens can be prevented would lower the risk for important types of cancer in many parts of the world.

ACKNOWLEDGMENTS

This research was supported in part by USPHS Grants CA-24217, CA-29602, and CA-30658, awarded by the National Cancer Institute, DHHS.

We wish to thank Mrs. Clara Horn for her assistance in editing and in preparing the manuscript.

REFERENCES

1. **Weisburger, J. H. and Williams, G. M.**, Chemical carcinogenesis, in *Cancer Medicine*, 2nd ed., Holland, J. F. and Frei, E., III, Eds., Lea & Febiger, Philadelphia, 1982, 42.
2. **Miller, E. C. and Miller, J. A.**, Mechanisms of chemical carcinogenesis, *Cancer*, 47, 1055, 1981.
3. **Weinberg, R. A.**, Oncogenes of spontaneous and chemically induced tumors, *Adv. Cancer Res.*, 36, 149, 1982.
4. **Nyce, J., Weinhouse, S., and Magee, P. N.**, 5-Methylcytosine depletion during tumour development: An extension of the miscoding concept, *Br. J. Cancer*, 48, 463, 1983.
5. **Ehrenberg, L., Brookes, P., Druckrey, H., Lagerlof, B., Litwin, J., and Williams, G. M.**, The relation of cancer induction and genetic damage, in *Evaluation of Genetic Risks of Environmental Chemicals*, Ambio Special Report No. 3, Ramel, C., Ed., Royal Swedish Academy of Science, Stockholm, Sweden, 1973, 15.

6. **Williams, G. M.**, Mechanisms of action of foodborne carcinogens, in *Environmental Aspects of Cancer: Role of Macro and Micronutrients*, Wynder, E. L., Leveille, G. A., Weisburger, J. H., and Livingston, G. E., Eds., Food and Nutrition Press, Westport, Conn., 1983, 83.

7. **Hecker, E., Fusenig, N. E., Kunz, W., Marks, F., and Thielmann, H. W., Eds.**, *Cocarcinogenesis and Biological Effects of Tumor Promoters*, Vol. 7, Raven Press, New York, 1982.

8. **Trosko, J. E., Yotti, L. P., Warren, S. T., Tsushimoto, and Chang, C-C.**, Inhibition of cell-cell communication by tumor promoters, in *Carcinogenesis and Biological Effects of Tumor Promoters*, Vol. 7, Hecker, E., Fusenig, N. E., Kunz, W., Marks, F., and Thielmann, H. W., Eds., Raven Press, New York, 1982, 565.

9. **Weinstein, I. B., Horowitz, A., Jeffrey, A., and Ivanovic, V.**, Cellular events in multistage carcinogenesis, in *Genes and Proteins in Oncogenesis*, Weinstein, I. B. and Vogel, H. J., Eds., Academic press, New York, 1983, 99.

10. **Slaga, T. J., Ed.**, International symposium on tumor promotion, *Environ. Health Perspect.*, 50, 3, 1983.

11. **Weisburger, J. H., Horn, C. L., and Barnes, W. S.**, Possible genotoxic carcinogens in foods in relation to cancer causation, *Semin. Oncol.*, 10, 330, 1983.

12. **Mirivish, S. S.**, The etiology of gastric cancer, *J. Natl. Cancer Inst.*, 71, 631, 1983.

13. **Hoffmann, D., Hecht, S. S., and Wynder, E. L.**, Tumor promoters and cocarcinogens in tobacco carcinogenesis, *Environ. Health Perspect.*, 50, 247, 1983.

14. **Wynder, E. L.**, Tumor enhancers: underestimated factors in the epidemiology of lifestyle-associated cancers, *Environ. Health Perspect.*, 50, 15, 1983.

15. **Reddy, B. S., Cohen, L. A., McCoy, G. D., Hill, P., Weisburger, J. H., and Wynder, E. L.**, Nutrition and its relationship to cancer, *Adv. Cancer Res.*, 32, 237, 1980.

16. **Weisburger, J. H., Wynder, E. L., with Horn, C. L.**, Nutritional factors and etiologic mechanisms in the causation of gastrointestinal cancers, *Cancer*, 50, 11, 1982.

17. **Wynder, E. L. and Cohen, L. A.**, A rationale for dietary intervention in the treatment of postmenopausal breast cancer patients, *Nutr. Cancer*, 3, 195, 1982.

18. **Doll, R. and Peto, R.**, The causes of cancer: quantitative estimates of avoidable risks of cancer in the United States today, *J. Natl. Cancer Inst.*, 66, 1191, 1981.

19. **Dennis, M. J., Massey, R. C., McWeeney, D. J., and Knowles, M. E.**, Analysis of polycyclic aromatic hydrocarbons in UK total diets, *Food Chem. Toxicol.*, 21, 569, 1983.

20. **Higginson, J.**, Etiology of gastrointestinal cancer in man, *NCI Monogr.*, 25, 191, 1967.

21. **Lijinsky, W. and Shubik, P.**, Polynuclear hydrocarbon carcinogens in cooked meat and smoked food, *Ind. Med. Surg.*, 34, 152, 1965.

22. **Howard, J. W. and Fazio, T.**, Review of polycyclic aromatic hydrocarbons in foods. Analytical methodology and reported findings of polycyclic aromatic hydrocarbons in foods, *J. Assoc. Off. Anal. Chem.*, 63, 1077, 1980.

23. **Stewart, H. L.**, Experimental alimentary tract cancer, *NCI Monogr.*, 25, 199, 1967.

24. **Homburger, F.**, Background data on tumor incidence in control animals (Syrian hamsters), *Prog. Exp. Tumor Res.*, 26, 259, 1983.

25. **Huggins, C. B.**, *Experimental Leukemia and Mammary Cancer*, University of Chicago Press, Chicago, 1979.

26. **Hecht, S. S., Grabowski, W., and Groth, K.**, Analysis of feces for benzo(a)pyrene after consumption of charcoal broiled beef by rats and humans, *Food Cosmet. Toxicol.*, 17, 223, 1979.

27. **Stich, H. F., Karim, J., Koropatnick, J., and Lo, L.**, Mutagenic action of ascorbic acid, *Nature (London)*, 260, 722, 1976.

28. **Norkus, E. P., Kuenzig, W., and Conney, A. H.**, Studies on the mutagenic activity of ascorbic acid in vitro and in vivo, *Mutat. Res.*, 117, 183, 1983.

29. **Ueno, I., Nakano, N., and Hirono, I.**, Metabolic fate of [^{14}C]quercetin in the ACI rat, *Jpn. J. Exp. Med.*, 53, 41, 1983.

30. **Takanashi, H., Shigetoshi, A., Hirono, I., Matsushima, T., and Sugimura, T.**, Carcinogenicity test of quercetin and kaempferol in rats by oral administration, *J. Food Safety*, 5, 55, 1983.

31. **Hatcher, J. F., Pamukcu, A. M., Erturk, E., and Bryan, G. T.**, Mutagenic and carcinogenic activities of quercetin, *Fed. Proc.*, 42, 786, 1983.

32. **Ertürk, E., Nunoya, T., Hatcher, J. F., Pamukcu, A. M., and Bryan, G. T.**, Comparison of bracken fern (BF) and quercetin (Q) carcinogenicity in rats (R), (Abstract 210), *Proc. AACR* 24, 53, 1983.

33. **Mori, H., Ohbayashi, F., Hirono, I., Shimada, T., and Williams, G. M.**, Absence of genotoxicity of the carcinogenic sulfated polysaccharides carrageenan and dextran sulfate in mammalian DNA repair and bacterial mutagenicity assays, *Nutr. Cancer*, 6, 92, 1984.

34. **Sugimura, T. and Sato, S.**, Mutagens-carcinogens in foods, *Cancer Res.*, 43(Suppl.), 2415s, 1983.

35. **Ray, S. and Ray, M.**, Formation of methylglyoxal from aminoacetone by amine oxidase from goat plasma, *J. Biol. Chem.*, 258, 3461, 1983.

36. **Sugimura, T.,** Mutagens, carcinogens, and tumor promoters in our daily food, *Cancer,* 49, 1970, 1982.
37. **Weisburger, J.,** Current views on mechanisms concerned with the etiology of cancers in the digestive tract, in *Pathophysiology of Carcinogenesis in Digestive Organs,* Farber, E., Ed., University of Tokyo Press, Tokyo, 1977, 1.
38. **Silverberg, E.,** Cancer statistics, *Ca-A Cancer J. Clin.,* 35, 19, 1985.
39. **Schottenfeld, D. and Fraumeni, J. F., Jr.,** *Cancer Epidemiology and Prevention,* W. B. Saunders, Philadelphia, 1982.
40. **Wynder, E. L., Peters, J. A., and Vivona, S.,** Symposium: nutrition in the causation of cancer, *Cancer Res.,* 35, 3235, 1975.
41. **Livingston, G. E., Moshy, R. J., and Chang, C. M., Eds.,** *The Role of Food Product Development in Implementing Dietary Guidelines,* Food and Nutrition Press, Westport, Conn., 1982.
42. **Wynder, E. L., Leveille, G. A., Weisburger, J. H., and Livingston, G. E., Eds.,** *Environmental Aspects of Cancer: The Role of Macro and Micro Components of Foods,* Food and Nutrition Press, Westport, Conn., 1983.
43. **Hirayama, T., Diet and Cancer,** *Nutr. and Cancer,* 1, 67—79,
44. **Weisburger, J. H. and Williams, G. M.,** Bioassay of carcinogens: in vitro and in vivo tests, in *Chemical Carcinogens,* Am. Chem. Soc. Monogr. 173, 2nd ed., Searle, C. E., Eds., American Chemical Society, Washington, D.C., 1984.
45. **Weisburger, J. H. and Williams, G. M.,** Carcinogen testing: current problems and new approaches, *Science,* 214, 401, 1981.
46. **Weisburger, J. H. and Fiala, E. S.,** Mechanisms of species, strain, dose effects in arylamine carcinogenesis, *NCI Monogr.,* 58, 41, 1981.
47. **Barnes, W. S., Lovelette, C. A., Tong, C., Williams, G. M., and Weisburger, J. H.,** Genotoxicity of the food mutagen 2-amino-3-methylimidazo[4,5-f]quinoline (IQ), and analogs, *Carcinogenesis,* 6, 441, 1985.
48. **Sugimura, T., Sato, S., and Takayama, S.,** New mutagenic heterocyclic amines found in amino acid and protein pyrolysates and in cooked food, in *Environmental Aspects of Cancer: The Role of Macro and Micro Components of Foods,* Wynder, E. L., Leveille, G. A, Weisburger, J. H., and Livingston, G. E., Eds., Food and Nutrition Press, Westport, Conn., 1983, 167.
49. **Nakayasu, M., Nakasoto, F., Sakamoto, H., Terada, M., and Sugimura, T.,** Mutagenic activity of heterocyclic amines in Chinese hamster lung cells with diphtheria toxin resistance as a marker, *Mutat. Res.,* 118, 91, 1983.
50. **Thompson, L. H., Carrano, A. V., Salazar, E., Felton, J. S., and Hatch, F. T.,** Comparative genotoxic effects of the cooked food related mutagens Trp-P-2 and IQ in bacteria and cultured mammalian cells, *Mutat. Res.,* 117, 243, 1983.
51. **Sasaki, M., Sugimura, K., Yoshida, M. A., and Kawachi, T.,** Chromosome aberrations and sister chromatid exchanges induced by tryptophan pyrolysates, Trp-P-1 and Trp-P-2 in cultured human and Chinese hamster cells, *Proc. Jpn. Acad.,* 56B, 332, 1980.
52. **Tsuda, H., Kato, K., Matsumoto, T., Yoshida, D., and Mizusaki, S.,** Chromosomal aberrations and morphological transformation of hamster embryonic cells induced by L-tryptophan pyrolyzate in vitro, *Mutat. Res.,* 49, 145, 1978.
53. **Tada, H., Tada, M., Sugawara, R., and Oikawa, A.,** Actions of amino-β-carbolines on induction of sister chromatid exchanges, *Mutat. Res.,* 116, 137, 1983.
54. **Tohada, H., Oikawa, A., Kawachi, T., and Sugimura, T.,** Induction of sister-chromatid exchanges by mutagens from amino acid and protein pyrolysates, *Mutat. Res.,* 77, 65, 1980.
55. **Imoue, K., Shibata, T., and Abe, T.,** Induction of sister-chromatid exchanges in human lymphocytes by indirect carcinogens with and without metabolic activation, *Mutat. Res.,* 117, 301, 1983.
56. **Matsukawa, N., Kawachi, T., Morino, K., Ohgaki, H., and Sugimura, T.,** Carcinogenicity in mice of mutagenic compounds from a tryptophan pyrolyzate, *Science,* 213, 346, 1981.
57. **Ohgaki, H., Kusama, K., Matsukura, N., Motino, K., Hasegawa, H., Sato, S., Takayama, S., Sugimura, S.,** Carcinogenicity in mice of a mutagenic compound, 2-amino-3-methylimidazo [4,5-f] quinoline from broiled sardine, cooked beef and beef extract, *Carcinogenesis,* 5, 921, 1984.
58. **Hosaka, S., Matsushima, T., Hirono, I., and Sugimura, T.,** Carcinogenic activity of 3-amino-1-methyl-5H-pyrido[4,3-b]-indole (Trp-P-2), a pyrolysis product of tryptophan, *Cancer Lett.,* 13, 23, 1981.
59. **Takayama, S., Nakatsuru, Y., Masuda, M., Ohgaki, H., Sato, S., and Sugimura, S.,** Demonstration of carcinogenicity in F344 rats of 2-amino-3-methylimidazo [4,5-f] quineline from broiled sardine, fried beef and beef extract, *Gann,* 75, 467, 1984.
60. **Barnes, W. S., Spingarn, N. E., Garvie-Gould, C., Vuolo, L. L., Wang, Y. Y., and Weisburger, J. H.,** Mutagens in cooked foods: possible consequences of the Maillard reaction in foods and nutrition, in *The Maillard Reaction in Foods and Nutrition,* ACS Symp. Series 215, Waller, G. R., and Feather, M. S., Eds., American Chemical Society, Washington, D. C., 1983, 485.

61. **Pariza, M. W., Loretz, L. J., Storkson, J. M., and Holland, N. C.,** Mutagens and modulator of mutagenesis in fried ground beef, *Cancer Res.,* 43, 2444, 1983.

62. **Nader, C. J., Spencer, L. K., and Weller, R. J.,** Mutagen production during pan-broiling compared with microwave irradiation of beef, *Cancer Lett.,* 13, 147, 1981.

63. **Matsumoto, T., Yoshida, D., and Tomita, H.,** Determination of mutagens, amino-α-carbolines in grilled foods and cigarette smoke condensate, *Cancer Lett.,* 12, 105, 1981.

64. **Sugimura, T., Kawachi, M., Nagao, M., Yamada, M., Takayama, S., Matsukawa, N., and Wakabayashi, K.,** in *Biochemical and Medical Aspects of Tryptophan Metabolism,* Elsevier/North Holland, Amsterdam, 1980, 297.

65. **Kasai, H., Yamaizumi, Z., Wakabayashi, K., Nagao, M., Sugimura, T., Yokoyama, S., Miyazawa, T., Spingarn, N. E., Weisburger, J. H., and Nishimura, S.,** Potent novel mutagens produced by broiling fish under normal conditions, *Proc. Jpn. Acad.,* 56B, 278, 1980.

66. **Yamaguchi, K., Shudo, K., Okamoto, T., Sugimura, T., and Kosuge, T.,** Presence of 2-aminopyrido[1,2α:3′,2′-d] imidazole in broiled cuttlefish, *Gann,* 71, 743, 1980.

67. **Sugimura, T., Nagao, M., and Wakabayashi, K.,** Mutagenic heterocyclic amines in cooked foods, in *Environmental Carcinogens Selected Methods of Analysis,* Vol. 40, Egan, H., Fishbein, L., Castegnaro, M., O'Neill, I. K., Bartsch, H., and Davis, W., Eds., International Agency for Research on Cancer, Lyon, France, 1981, 251.

68. **Kasai, H., Yamaizumi, Z., Shiomi, T., Yokoyama, S., Miyazawa, T., Wakabayashi, K., Nagao, M., Sugimura, T., and Nishimura, S.,** Structure of a potent mutagen isolated from fried beef, *Chem. Lett.,* p. 485, 1981.

69. **Bjeldanes, L. F. and Chang, G. W.,** Mutagenic activity of quercetin and related compounds, *Science,* 197, 577, 1977.

70. **Takahashi, Y., Nagao, M., Fujino, T., Yamaizumi, Z., and Sugimura, T.,** Mutagens in Japanese pickle identified as flavonoids. *Mutat. Res.,* 68, 117, 1979.

71. **Bjeldanes, L. F., Morris, M. M., Felton, J. S., Healy, S., Stuerman, D., Berry, P., Timourian, H.,and Hatch, F. T., III,** Survey by Ames/Salmonella test of mutagen formation in secondary sources of cooked dietary protein, *Food Chem. Toxicol.,* 20, 365, 1982.

72. **Waller, G. R. and Feather, M. S., Eds.,** *The Maillard Reaction in Foods and Nutrition.* ACS Symp. Series 215, American Chemical Society, Washington, D. C., 1983.

73. **Spingarn, N. E., Garvie-Gould, C. T., and Slocum, L. A.,** Formation of mutagens in sugar-amino acid model systems, *J. Agric. Food Chem.,* 31, 301, 1983.

74. **Shibamoto, T., Nishimura, O., and Mihara, S.,** Mutagenicity of products obtained from a maltol-ammonia browning model system, *J. Agric. Food Chem.,* 29, 643, 1981.

75. **Powrie, W. D., Wu, C. H., Rosin, M. P., and Stich, H. F.,** Clastogenic and mutagenic activities of Maillard reaction model systems, *J. Food Sci.,* 46, 1433, 1981.

76. **Aeschbacher, H. U., Chappius, C. H.,Managanel, M., and Aeschbach, R.,** Investigation of Maillard products in bacterial mutagenicity test systems, *Prog. Food Nutr. Sci.,* 5(1—6), 279, 1981.

77. **Matsushima, T.,** Mechanisms of conversion of food components to mutagens and carcinogens, in, *Molecular Interrelations of Nutrition and Cancer,* Arnott, M. S., van Eys, J., and Wang, Y-M, Eds., Raven Press, New York, 1982,35.

78. **Jägerstad, M., Olssson, K., Grivas, S., Negishi, C., Wakabayashi, K., Tsuda, M., Sato, S., and Sugimura, T.,** Formation of 2-amino-3,8-dimethylimidazo[4,5-f]quinoxaline in a model system by heating creatinine, glycine, and glucose, *Mutat. Res.,* 126, 239, 1984.

79. **Tanaka, T., Barnes, W. S., Weisburger, J. H., and Williams, G. M.,** Multipotential carcinogenicity of the fried food mutagen 2-amino-3-methylimidazo [4-5f] quinoline in rats, *Jpn. J. Cancer Res. (Gann),* 76, 570, 1985.

80. **Kato, R.,** In vitro metabolism of heterocyclic amines, presented at U.S.-Japan Cooperative Cancer Research Program Workshop, Honolulu, Hawaii, February 17 to 18, 1984.

81. **Ishikawa, T., Takayama, S., Kitagawa, T., et al.,** In vivo experiments on tryptophan pyrolysis products, in *Naturally Occurring Carcinogens-Mutagens and Modulators of Carcinogenesis,* Japan Science Society Press, Baltimore, 1979, 159.

87. **Shudo, K.,** Organic chemistry and formation of DNA adducts by heterocyclic amines, presented at U.S.-Japan Cooperative Cancer Research Program Workshop, Honolulu, Hawaii, February 17 to 18, 1984.

88. **Spingarn, N. E., Garvie-Gould, C., Vuolo, L. L., Weisburger, J. H.,** Formation of mutagens in cooked foods. IV. Effect of fat content in fried beef patties, *Cancer Lett.,* 12, 93, 1981.

89. **Barnes, W. S. and Weisburger, J. H.,** Formation of mutagens in cooked foods. VI. Modulation of mutagen formation by iron and ethylenediaminetetraacetic acid (EDTA) in fried beef, *Cancer Lett.,* 24, 221, 1984.

90. **Taylor, R. T.,** Mutagen formation in a model beef boiling system. presented at U.S.-Japan Cooperative Cancer Research Program Workshop, Honolulu, Hawaii, February 17 to 18, 1984.

91. **Hayatsu, H., Inoue, K., Ohta, H., Namba, T., Tagawa, K., Hayatsu, T., Makita, M., and Watoya, Y.,** Inhibition of the mutagenicity of cooked-beef basic fraction by its acidic fraction, *Mutat. Res.,* 91, 437, 1981.

92. **Kada, T., Morita, K., and Inoue, T.,** Anti-mutagenic action of vegetable factor(s) on the mutagenic principle of tryptophan pyrolysate, *Mutat. Res.,* 53, 351, 1978.

93. **Lai, C., Butler, M. A., and Matney, T. S.,** Antimutagenic activities of common vegetables and their chlorophyll content, *Mutat. Res.,* 77, 245, 1980.

94. **Wynder, E. L. and Reddy, B. S.,** Dietary fat and fiber and colon cancer, *Semin. Oncol.,* 10, 320, 1984.

95. **Rose, D. P., Ed.,** *Endocrinology of Cancer,* Vol. 1 to 3, CRC Press, Boca Raton, Fla., 1982.

96. **Winawer, S., Schottenfeld, D., and Sherlock, P., Eds.,** *Colorectal Cancer: Prevention, Epidemiology, and Screening,* Raven Press, New York, 1980.

97. **Newell, G. R. and Ellison, N. M.,** *Nutrition and Cancer: Etiology and Treatment,* Raven Press, New York, 1981.

98. **Arnott, M. S., van Eys, J., and Wang, Y-M., Eds.,** *Molecular Interrelations of Nutrition and Cancer,* Raven Press, New York, 1982.

99. **Weisburger, J. H., Reddy, B. S., Barnes, W., and Wynder, E. L.,** Bile acids, but not neutral sterols are tumor promoters in the colon in man and in rodents, *Environ. Health Perspect.,* 50, 101, 1983.

100. **Wynder, E. L.,** Reflections on diet, nutrition, and cancer, *Cancer Res.,* 43, 3024, 1983.

101. **Magee, P. N., Ed.,** *Nitrosamines and Human Cancer,* Banbury Rep. 12, Cold Spring Harbor Laboratories, Cold Spring Harbor, N. Y., 1982.

102. **Scanlan, R. A. and Tannenbaum, S. R., Eds.,** *N-Nitroso Compounds,* ACS Symp. Series 174, American Chemical Society, Washington, D. C., 1981.

103. **Hill, M. J.,** Environmental and genetic factors in gastrointestinal cancer, in, *Precancerous Lesions of the Gastrointestinal Tract,* Sherlock, P., Morson, B. C., Barbara, L., and Veronesi, U., Eds., Raven Press, New York, 1983, 1.

104. **Joossens, J. V. and Geboers, J.,** Epidemiology of gastric cancer: a clue to etiology, in *Precancerous Lesions of the Gastro-intestinal Tract,* Sherlock, P., Morson, B. C., Barbara, L., and Veronesi, U., Eds., Raven Press, New York, 1983, 97.

105. **Correa, P., Cuello, C., Fajardo, L. F., Haenszel, W., Bolanos, O., and de Ramirez, B.,** Diet and gastric cancer: nutrition survey in a high-risk area, *J. Natl. Cancer Inst.,* 70, 673, 1983.

106. **Bjelke, E.,** The recession of stomach cancer: selected aspects, in *Trends in Cancer Incidence,* Magnus, K., Ed., Hemisphere Publishing, New York, 1982, 165.

107. **Weisburger, J. H., Marquardt, H., Mower, H. F., Hirota, N., Mori, H., and Williams, G. M.,** Inhibition of carcinogenesis: vitamin C and the prevention of gastric cancer, *Prev. Med.,* 9, 352, 1980.

108. **Armijo, R., Gonzalez, A., Orellana, M., Coulson, A. H., Sayre, J. W., and Detels, R.,** Epidemiology of gastric cancer in Chile. II. Nitrate exposures and stomach cancer frequency, *Int. J. Epidemiol.,* 10, 57, 1981.

109. **Archer, M. C.,** Hazards of nitrate, nitrite, and N-nitroso compounds in human nutrition, *Nutr. Toxicol.,* 1, 327, 1982.

110. **Hartman, P. E.,** Nitrate and nitrite ingestion and gastric cancer mortality, *Environ. Mutagenesis,* 5, 111, 1983.

111. **Ichinotsubo, D. Y. and Mower, H. F.,** Mutagens in dried/salted Hawaiian fish, *J. Agric. Food Chem.,* 30, 937, 1982.

112. **Stemmermann, G. N.,** Gastric cancer in the Hawaii Japanese, *Gann,* 68, 525, 1977.

113. **Sugimura, T. and Kawachi, T.,** Experimental stomach carcinogenesis, in *Gastro-Intestinal Tract Cancer,* Lipkin, M. and Good, R., Eds., Plenum Press, New York, 1978, 327.

114. **Sander, J., Schweinsberg, F., LaBar, J., Bürkle, G., and Schweinsberg, E.,** Nitrite and nitrosable amino compounds in carcinogenesis, *Gann Monogr.,* 17, 145, 1975.

115. **Ochiai, M., Wakabayashi, K., Nagao, M., and Sugimura, T.,** Tyramine is a major mutagen precursor in soy sauce, being convertible to a mutagen by nitrite, *Gann,* 75, 1, 1984.

116. **Piacek-Llanes, B. G. and Tannenbaum, S. R.,** Formation of an activated N-nitroso compound in nitrite-treated fava beans *(Vicia faba), Carcinogenesis,* 3, 1379, 1982.

117. **Yano, K.,** Alkylating activity of processed fish products treated with sodium nitrite in simulated gastric juice, *Gann,* 72, 451, 1981.

118. **Boyd, S. C., Sasame, H. A., and Boyd, M. R.,** High concentrations of glutathione in glandular stomach: possible implications for carcinogenesis, *Science,* 205, 1010, 1979.

119. **Kleihues, P. and Wiestler, O.,** Involvement of thiols in MNNG-induced gastric cancer: biochemical and autoradiographic studies, *IARC Sci. Publ.,* 57, 603, 1984. Effects, IARC Sci. Publ., in press, 1983.

120. **Toth, B., Nagel, D., and Ross, A.,** Gastric tumorigenesis by a single dose of 4-(hydroxy-methyl)benzenediazonium ion of *Agaricus bisporus, Br. J. Cancer,* 46, 417, 1982.

121. **Wakabayashi, K., Ochiai, M., Saito, H., Tsuda, M., Suwa, Y., Nagao, M., and Sugimura, T.,** Presence of 1-methyl-1,2,3,4-tetrahydro-β-carboline-3-carboxylic acid, a precursor of a mutagenic nitroso compound, in soy sauce, *Proc. Natl. Acad. Sci. U.S.A.*, 80, 2912, 1983.
122. **Takahashi, M., Kokubo, T., Furukawa, F., Kurokawa, Y., Tatematsu, M., and Hayashi, Y.,** Effect of high salt diet on rat gastric carcinogenesis induced by N-methyl-N'-nitro-nitrosoguanidine, *Gann*, 74, 28, 1983.
123. **Shirai, T., Imaida, K., Fukushima, S., Hasegawa, R., Tatematsu, M., and Ito, N.,** Effects of NaCl, Tween 60 and a low dose of N-ethyl-N'-nitro-N-nitrosoguanidine on gastric carcinogenesis of rat given a single dose of N-methyl-N'-nitro-N-nitrosoguanidine, *Carcinogenesis*, 3(12), 1419, 1982.
124. **Kono, S., Ideda, M., and Ogata, M.,** Salt and geographical mortality of gastric cancer and stroke in Japan, *J. Epidemiol. Commun. Health*, 37, 43, 1983.
125. **Correa, P., Haenszel, W., and Tannenbaum, S.,** Epidemiology of gastric carcinoma: review and future prospects, *NCI Monogr.*, 62, 129, 1983.
126. **Stähelin, H. B., Buess, E., Rösel, F., Widmer, L. K., and Brubacher, G.,** Vitamin A, cardiovascular risk factors, and mortality, *Lancet*, 1, 394, 1982.
127. **Sporn, M. B. and Roberts, A. B.,** Role of retinoids in differentiation and carcinogenesis, *Cancer Res.*, 43, 3034, 1983.
128. **Tuyns, A. J., Pequignot, G., and Jensen, D. M.,** Role of diet, alcohol and tobacco in oesophageal cancer, as illustrated by two contrasting high-incidence areas in the north of Iran and west of France, in *Frontiers of Gastrointestinal Research*, Vol. 4, van der Reis, L., Ed., S. Karger, Basel, 1979, 101.
129. Joint Iran-International Agency for Research on Cancer Study Group, Esophageal cancer studies in the Caspian Littoral of Iran: results of population studies — a prodrome, *J. Natl. Cancer Inst.*, 59, 1127, 1977.
130. **Malaveille, C., Friesen, M., Camus, A-M., Garren, L., Hautefeuille, A., Bereziat, J. C., Ghadirian, P., Day, N. E., and Bartsch, H.,** Mutagens produced by the pyrolysis of opium and its alkaloids as possible risk factors in cancer of the bladder and oesophagus, *Carcinogenesis*, 3, 577, 1982.
131. **Cheng, S-J., Sala, M., Li, M. H., Wang, M-Y., Pot-Deprun, J., and Chouroulinkov, J.,** Mutagenic, transforming and promoting effect of pickled vegetables from Linxian county, China, *Carcinogenesis*, 1, 685, 1980.
132. **Huang, D. P., Ho, J. H. C., Webb, K. S., Wood, B. J., and Gough, T. A.,** Volatile nitrosamines in salt-preserved fish before and after cooking, *Food Cosmet. Toxicol.*, 19, 167, 1981.
133. **Lu, S.H., Camus, A. M., Ji, C., Wang, Y. L., Wang, M. Y., and Bartsch, H.,** Mutagenicity in *Salmonella typhimurium* of N-3-methylbutyl-N-1-methyl-acetonyl-nitrosamine and N-methyl-N-benzylnitrosamine, N-nitrosation products isolated from corn-bread contaminated with commonly occurring moulds in Linshien county, a high incidence area for oesophageal cancer in Northern China, *Carcinogenesis*, 1, 867, 1980.
134. **Armstrong, R. W., Armstrong, M. J., Yu, M. C., and Henderson, B. E.,** Salted fish and inhalants as risk factors for nasopharyngeal carcinoma in Malaysian Chinese, *Cancer Res.*, 43, 2967, 1983.
135. **Lu, S. H. and Lin, P.,** Recent research on the etiology of esophageal cancer in China, *Z. Gastroenterol.*, 20, 361, 1982.
136. **Fong, L. Y., Ho, J. H., and Huang, D. P.,** Preserved foods as possible cancer hazards: WA rats fed salted fish have mutagenic urine, *Int. J. Cancer*, 23, 542, 1979.
137. **Lu, S-H., Camus, A-M., Tomatis, L., and Bartsch, H.,** Mutagenicity of extracts of pickled vegetables collected in Linshien County, a high-incidence area for esophageal cancer in Northern China, *J. Natl. Cancer Inst.*, 66, 33, 1981.
138. **Gupta, I., Baptista, J., Bruce, W. R., Che, C. T., Furrer, R., Gingerich, J. S., Grey, A. A., Marai, L., Yates, P., and Krepinsky, J. J.,** Structures of fecapentaenes, the mutagens of bacterial origin isolated from human feces, *Biochemistry*, 22, 241, 1983.
139. **Ehrich, M. J. E., Aswell, J. E., van Tassell, R. L., and Wilkins, T. D.,** Mutagens in the feces of 3 South African populations at different levels of risk for colon cancer, *Mutat. Res.*, 64, 231, 1979.
140. **Van Tassell, R. L., MacDonald, D. K., and Wilkins, T. D.,** Stimulation of mutagen production in human feces by bile and bile acids, *Mutat. Res.*, 103, 233, 1982.
141. **Reddy, B. S., Sharma, C., Darby, L., Laakso, K., and Wynder, E. L.,** Metabolic epidemiology of large bowel cancer. Fecal mutagens in high- and low-risk populations for colon cancer. A preliminary report, *Mutat. Res.*, 72, 511, 1980.
142. **Mower, H. F., Ichinotsubo, D., Wang, L. W., Mandel, G., Stemmermann, G., Nomura, A., Heilbrun, L., Kamiyama, S., and Shimada, A.,** Fecal mutagens in two Japanese populations with different colon cancer risks, *Cancer Res.*, 42, 1164, 1982.
143. **Venitt, S. and Bosworth, D.,** The development of anaerobic methods for bacterial mutation assays: aerobic and anaerobic fluctuation tests of human faecal extracts and reference mutagens, *Carcinogenesis*, 4, 339 1983.
144. **Hirai, N., Kingston, D. G. I., Van Tassell, R. L., and Wilkins, T. D.,** Structure elucidation of a potent mutagen from human feces, *J. Am. Chem. Soc.*, 104, 6149, 1982.

Chapter 9

FORMATION AND OCCURRENCE OF NITROSAMINES IN FOOD

Nrisinha P. Sen

TABLE OF CONTENTS

I. INTRODUCTION

Man is exposed daily to a large number of chemicals (both man-made and of natural origin) through consumption of foods and beverages. Although the majority of these chemicals (e.g., amino acids, minerals, vitamins) are essential for proper nutrition and growth of an individual, many are undesirable and some are extremely toxic. Again all that is natural is not safe; many naturally occurring compounds (e.g., aflatoxins, cycasin, oxalates, thiocyanates, arsenic) are extremely toxic.[1] In some cases the toxic constituents of the foodstuff originate from secondary sources such as residues from packaging materials or are formed during processing, storage or cooking of foods. The formation of *N*-nitrosamines (simply called nitrosamines) in foods can be best described as falling into the last category. These compounds are usually formed as a result of processing or cooking of foods although their occurrence from other sources (e.g., natural occurrence of streptozotocin) can not be completely ruled out. Since the precursors (amines and nitrite) of nitrosamines are ubiquitous in nature or are widely used in industry it is not surprising that traces of various nitrosamines have been reported to occur in a wide variety of foods and beverages, and much higher levels (up to 1000 ppm) have been detected in tobacco products, cosmetics, cutting oils, pesticide formulations, and rubber products.[2] Our concern in this area mainly stems from the fact that most nitrosamines are potent carcinogens and, therefore, their presence in foods or other environmental media can be regarded as a potential hazard to human health.

There are various *N*-nitroso compounds. The most common ones are the nitrosamines such as *N*-nitrosodimethylamine (NDMA), *N*-nitrosodiethylamine (NDEA) and *N*-nitrosopyrrolidine (NPYR). Other examples of *N*-nitroso compounds include the nitrosamino acids, nitrosamides, nitrosoureas, nitrosoguanidines, nitrosourethanes, and nitrosocyanamides. Of nearly 300 *N*-nitroso compounds tested thus far over 80% have been shown to be carcinogenic in various species of animals.[3,4] Most of them are also mutagenic and some have been shown to be teratogenic.[4] Therefore, the general concern due to the finding of nitrosamines in various foodstuffs is understandable. Moreover, the occurrence of nitrosamines in foods is not the only thing that is of concern. The possibility of formation, from ingested nitrite and amines in foods, of these carcinogens in vivo in the acidic environment of human stomach adds another dimension to the problem. If one has to assess the role of nitrosamines in foods on human health both the exogenous and endogenous sources should be taken into account. A comprehensive review of the subject is beyond the scope of this mini-review. Therefore, this article will present brief summaries of the major findings and concentrate on some of the more recent results.

II. CHEMISTRY OF FORMATION OF *N*-NITROSO COMPOUNDS

A. Nitrosation in Aqueous Solution

The chemistry and formation of *N*-nitroso compounds have been studied extensively and reviewed by Ridd, Mirvish, and others.[5-9] In general, nitrosamines are prepared by treating aqueous solutions of secondary amines with nitrous acid under acidic conditions. The nitrosation can also be carried out in organic solvents in the presence of a suitable nitrosating agent such as NOCl, N_2O_3, or $NOBF_4$. By itself the nitrous acid or the nitrite ion is a poor nitrosating agent, but under moderately acidic conditions it forms (Equation 1) nitrous anhydride which is a strong nitrosating agent. Alternatively, in the presence of a nucleophilic anion (e.g., SCN^-, I^-) it can be converted to a more active nitrosating species, NOX, X^- being the nucleophilic anion (Equation 2). Thus, the overall nitrosation of a dialkylamine in aqueous solution can be represented by Equations 1 to 3:

$$2 \text{ HONO} \rightleftharpoons \text{O=N-O-N=O} + H_2O \qquad (1)$$

$$HONO + X^- + H^+ \rightleftharpoons NOX + H_2O \qquad (2)$$

$$\begin{array}{c} R \\ \diagdown \\ NH + \\ \diagup \\ R \end{array} \begin{array}{c} O{=}N{-}O{-}N{=}O \\ or \\ NOX \end{array} \rightleftharpoons \begin{array}{c} R \\ \diagdown \\ N{-}N{=}O + \\ \diagup \\ R \end{array} \begin{array}{c} NO_2^- \\ or \\ X^- \end{array} + H^+$$

Dialkylamine   Dialkylnitrosamine   (3)

Under strongly acidic (pH < 2) conditions more powerful nitrosting agents such as nitrous acidium ion (H_2ONO^+) or nitrosonium ion (NO^+) are formed as shown in Equations 4 and 5:

$$HONO + H^+ \rightleftharpoons H_2ONO^+ \qquad (4)$$

$$H_2ONO^+ \rightleftharpoons NO^+ + H_2O \qquad (5)$$

Since the nitrosation of weakly basic amines (e.g., *N*-methylaniline) and amides are more favorable under strongly acidic conditions, H_2ONO^+ or NO^+ probably play a more active role in the nitrosation of these compounds. Although nitrosation under such strongly acidic conditions is not so relevant to the situations existing with foods, the mechanism may be applicable to *N*-nitrosation occurring in vivo in the acidic environment of the human stomach.

The amounts of *N*-nitroso compounds that will be formed in a food system or in vivo in the human stomach will depend on a number of factors such as the kinetics (rate constants) of the nitrosation reaction, concentration and nature of both the amine and of the nitrosating species, pH, temperature, and the presence of competing reactions (e.g., catalysts or inhibitors). Since two molecules of nitrous acid are needed to form one molecule of nitrous anhydride (Equation 1), which is the main nitrosating agent under moderately acidic conditions, the rate of nitrosation of a secondary amine is proportional to the square of the nitrite concentration:

$$Rate = k_1[R_2NH] [HNO_2]^2 \qquad (6)$$

or

$$Rate = k_2[total\ amine] [nitrite]^2 \qquad (7)$$

where k_1 and k_2 are the respective rate constants, and $[R_2NH]$ and [total amine] represent the concentrations of the unprotonated amine and the total (protonated + unprotonated) amine, respectively. The rate constant k_1 (Equation 6), is independent of pH whereas k_2 (Equation 7) varies with pH and has been shown to have a maximum value at pH 3.4, the pKa of nitrous acid. It should be clearly evident from the above equations that a gradual lowering (say from pH 9) of pH of the reaction medium will have opposing effects. On the one hand it would enhance the formation of nitrous anhydride (favored by acidic pH) but, on the other, it would increase the protonation of R_2NH thus decreasing the concentration of the unprotonated amine. In the pH range of 9 to 5 the nitrosation rate of dimethylamine, a commonly occurring secondary amine in foods, would be expected to increase by a factor of 10 for each decrease of 1 unit of pH.[6] Mirvish has extensively studied the kinetics of nitrosation of various amines.[7] A simplified version of his data is presented in Table 1. The most important observation that one can make is that the nitrosation rate is inversely proportional to the basicity of the amine. Or, in other words, a weakly basic amine (e.g., *N*-methylaniline) is nitrosated much faster than a strongly basic amine like dimethylamine.

In contrast to the secondary amines, tertiary amines, with a few exceptions (e.g., ami-

Table 1
RELATIVE RATES OF NITROSATION OF VARIOUS SECONDARY AMINES

Amine	pKa	Optimum pH of nitrosation	Rate constant k_2 (Equation 7)[a]	Relative rate
Dimethylamine	10.72	3.4	0.0017	1
Piperidine	11.2	3.0	0.00045	0.26
Pyrrolidine	11.27	3.0	0.0053	3.1
Proline	—	2.5	0.037	21.8
Hydroxyproline	—	2.5	0.31	182.3
Sarcosine	—	2.5	0.23	135.3
Prolylglycine	8.97	3.0	0.25	147
Morpholine	8.7	3.4	0.42	247
N-Methylbenzylamine	9.54	3.0	0.013	7.6
N-Methylaniline	4.85	—	250	147,058

[a]   Values at the optimum pH and 25°C (except at pH 1 and 0°C for N-methylaniline), in M^{-2} sec^{-1} (Reference 1, p.4 to 11).

From Mirvish, S., *Toxicol. Appl. Pharmacol.*, 31, 325, 1975. With permission.

nopyrene, gramine), nitrosate very slowly. Mirvish studied the nitrosation of trimethylamine and found the rate constant k_2 at 25°C to be only 2.4×10^{-7} M^{-2} sec^{-1} which is approximately 10,000 times less than that for dimethylamine.[7] Although trimethylamine is present in some foods, especially in marine fish, in high concentrations such a slow rate of nitrosation for this compound suggests that it would not lead to any significant formation of NDMA. Significant formation of NDMA can be observed only in the presence of high concentration of nitrite and at an elevated temperature. Smith and Loeppky have proposed a mechanism of nitrosation of tertiary amines that involves preliminary nitrosative dealkylation of the tertiary amine to form a secondary amine which is then nitrosated to a N-nitrosamine in the usual manner.[11]

In general, quaternary amino compounds nitrosate extremely slowly.[12] Ohshima and Kawabata have studied the mechanism of nitrosation of trimethylamine oxide (TMAO), a quaternary amine, which is highly abundant in marine fish and some seafoods.[13,14] The nitrosation rate of TMAO at 100°C (pH 3.0) was found to be proportional to the cube of the nitrite concentration but at 25°C it was proportional to the nitrite concentration. They proposed a dual mechanism of nitrosation to explain the above observation.[13]

As can be seen from Table 1 the nitrosation rates for the amino acids and of the dipeptide prolylglycine bearing a secondary amino group are much faster than that for the strongly basic amines. Their pH optima for nitrosation are also slightly lower — more close to that existing in the human stomach. Since many of these compounds are abundant in foods, the formation of the corresponding nitroso derivatives in the acidic environment of the human stomach (if enough nitrite is present) is a real possibilty.[14,15] As will be discussed later this has been shown to be the case. More recently, Ohshima et al. studied the kinetics of nitrosation of thiazolidine-4-carboxylic acid (TCA) and 2-methyl-thiazolidine-4-carboxylic acid (MTCA).[16] The nitrosation rate of TCA was found to increase linearly with decreasing pH in the range of pH 1 to 6, whereas that for MTCA was found to be maximum at pH 3.5. The two compounds were, respectively, found to nitrosate approximately 400 and 150 times faster than proline.

Thus it is clear that the formation (both in vitro and in vivo) of only those amino compounds that nitrosate rapidly (e.g., morpholine, N-methylaniline, sarcosine, proline, TCA) or are present in foods in high concentrations (e.g., dimethyl- and trimethylamines, proline) should be of interest. Fortunately, the former type (weakly basic) of amines either occur rarely in foods, or are present only in trace levels, or, if present in high concentrations (e.g., proline), the corresponding nitroso derivative (i.e., nitrosoproline) is noncarcinogenic.[4] This statement may not be true for TCA and MTCA since very little is known about the toxicity of the nopyrene, gramine), nitrosate very slowly. Mirvish studied the nitrosation of trimethylamine and found the rate constant k_2 at 25°C to be only 2.4×10^{-7} M^{-2} sec^{-1} which is approximately 10,000 times less than that for dimethylamine.[7] Although trimethylamine is present in some foods, especially in marine fish, in high concentrations such a slow rate of nitrosation for this compound suggests that it would not lead to any significant formation of NDMA. corresponding nitroso derivatives. Similarly, very little information exists with respect to the formation and carcinogenicity of nitrosated peptides.

The nitrosation of secondary amides including ureas, guanidines, and urethanes proceeds rapidly in aqueous acidic solution.[7] The reaction rate follows Equations 8 and 9:

$$HONO + H^+ \rightleftharpoons H_2ONO^+ \tag{4}$$

$$RNH \cdot COR' + H_2ONO^+ \rightleftharpoons RN(NO) \cdot COR' + H_3O^+ \tag{8}$$

$$Rate = k_3[RNH \cdot COR'] [HNO_2] [H^+] \tag{9}$$

In contrast to the nitrosation of secondary amines the above rate equation is first order with respect to $[HNO_2]$, and the rate increases about tenfold for each 1 unit decrease in pH without showing a pH optimum.[7] Since the nitrosation of these weakly basic amides proceeds best in strongly acidic solution the main nitrosating species is nitrous acidium ion and not nitrous anhydride. Mirvish studied the nitrosation of 21 amides which included several naturally occurring amides such as dihydrothymine, methylguanidine, hydantoic acid, DL-citrulline, and ethylcyanamide.[7] As a general rule the rate of nitrosation was found to be the lowest for simple alkyl- and arylamides and highest for ethylene urea. N-alkylureas and N-alkylcarbamates were found to nitrosate at an intermediate rate. Since the last two categories of amides nitrosate very rapidly their occurrence in food in moderate-to-high concentrations should be viewed with concern.

B. Nitrosation by Nitrogen Oxide Gases (NO$_x$)

Thus far, the discussion has been restricted to nitrosation by nitrous acid or related nitrosating agents produced under aqueous acidic conditions. Various studies have shown that gaseous nitrogen oxides can also directly nitrosate a secondary amino group.[17,18] Among the various oxides, NO_2, N_2O_4, and N_2O_3 can nitrosate a secondary amine very rapidly both in the lipid phase as well as in aqueous solutions. The nitrosation rate in the lipid phase is considerably faster than in the aqueous phase. In addition to the formation of N-nitrosamines nitrosation of secondary amines by N_2O_4 also results in the formation of some N-nitramines. In the absence of oxygen nitric oxide (NO) is a poor nitrosating agent, but it has been reported to nitrosate secondary amines under anaerobic conditions in the presence of iodine or several trace metal ions (Cu^{++}, Cu$^+$, Fe^{++}, etc.).[18,19] For a detailed discussion of nitrosation by NO$_x$ gases see the articles by Challis and co-workers.[17-19]

NO$_x$ gases also play an important role in the formation of nitrosamines in foods. The formation of NDMA in fish meal, malt, and skim milk powder as a result of direct drying by hot flue gases, which contain NO$_x$ gases, can be cited as examples.[20-22] NO$_x$ gases present in wood smoke or that formed during combustion of propane (gas stove) can nitrosate amines in foods during smoking or cooking.[23,24] As will be discussed in Section IV.B, there is considerable evidence to support the conclusion that nitrogen oxides also play a significant role in the formating of NPYR during frying of bacon.

C. Nitrosation of Primary Amines

The reaction of primary amines with nitrous acid is well known to the students of organic

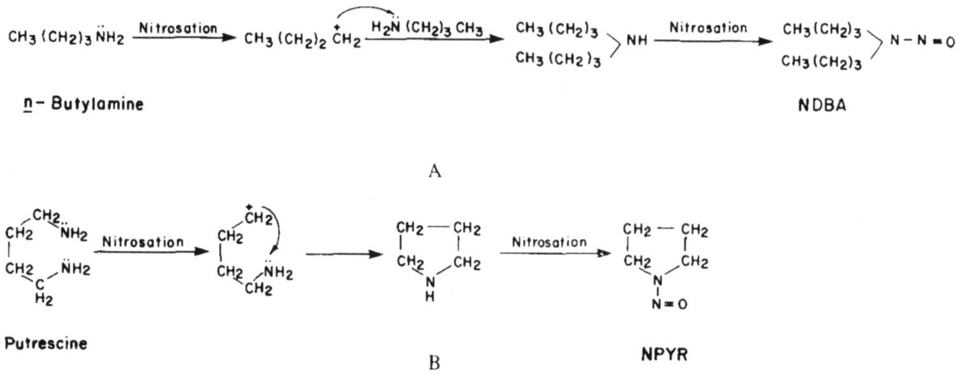

FIGURE 1.   Proposed pathways for the formation of (A) NDBA and (B) NPYR after nitrosation of butylamine and putrescine, respectively. Adapted from Scanlan.[28]

chemistry.[5,25] It proceeds vigorously with the evolution of nitrogen gas and subsequent formation of alcohols, olefins, ethers, or alkyl halides depending on the conditions of the reaction (Equation 10). As in the case of the secondary amines the first step in the nitrosation of both aliphatic and aromatic primary amines involves formation of a *N*-nitroso derivative. But because of the presence of a hydrogen on the amino nitrogen, the primary amine *N*-nitroso intermediate immediately tautomerizes and dehydrates to a diazonium ion.[25] An aryldiazonium ion thus produced is relatively stable at 0°C but the aliphatic counterpart is highly unstable and decomposes to nitrogen and a carbonium ion which then can react with various nucleophiles to give substitution, elimination, and rearrangement products as mentioned above.

$$R\text{–}\ddot{N}H_2 \xrightarrow[-H_2O]{HNO_2} R\text{–}\overset{H}{N}\text{–}N{=}O \rightarrow R\text{–}\ddot{N}{=}N\text{–}OH \xrightarrow[-H_2O]{H^+} R\text{–}N^+{\equiv}N\text{:}$$

$$\xrightarrow{-N_2} R^+ \xrightarrow{\text{Nucleophiles}} \begin{array}{l}\text{Olefins,}\\ \text{alcohols,}\\ \text{etc.}\end{array} \tag{10}$$

The nitrogen atom (containing an unpaired electron pair) of an unreacted primary amine in the same solution can also act as a nucleophile and react with the carbonium ion to produce a secondary amine.[1,5,26] Similarly, an aliphatic diamine (e.g., putrescine) can lead to the formation of a heterocylic secondary amine after nitrosation and cyclization.[26,27] The proposed pathways for the formation of nitrosamines from a typical aliphatic primary mono- and a diamine are shown in Figure 1.[28] Because of the many steps involved and other side reactions the yield of nitrosamine by such pathways is usually very low. Warthesen et al. reported only 0.01% yield of *N*-nitrosodi-*n*-butylamine (NDBA) from nitrosation of n-butylamine and 2.8 to 9.2% yields of NPYR after nitrosation of putrescine in the presence of high reagent concentrations.[26] Therefore, it is highly unlikely that nitrosation of primary amines would lead to any significant formation of nitrosamines either in vivo or in foods.

Hildrum et al. also studied the nitrosation of spermine and spermidine polyamines which contain both primary and secondary amine groups.[27] In each case the products were found to be very complex and consisted of several *N*-nitrosamines. For example, nitrosation of spermidine yielded NPYR, γ-butenyl (β-propenyl) nitrosamine, γ-butenyl (γ-propanol) nitrosamine, δ-butanol (β-propenyl) nitrosamine, and δ-butylchloride-(β-propenyl) nitrosamine. Although both spermine and spermidine are reported to occur in animal tissues no data are available on the occurrence or formation of the above nitrosamines (except NPYR) in nitrite-treated meats.

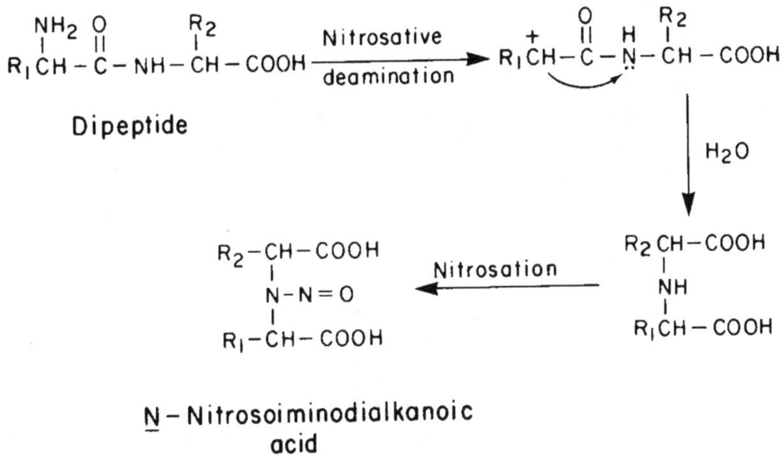

FIGURE 2. Mechanism of formation of *N*-nitrosoiminodialkanoic acids from peptides. (Adapted from Outram, J. R. and Pollock, J. R. A., Production of *N*-Nitrosoiminodialkanoic Acids by Nitrite in Gastric Juice, Publ. No. 57, International Agency for Research on Cancer, Lyon, 1984, 71.)

D. Nitrosation of Peptides

The possibility of nitrosation of the $-NH$ group in peptide linkages in proteins and smaller peptides has been discussed for many years and has been investigated recently by several laboratories.[29-31] But because of the extreme instability of the compounds the progress has been slow. Since most nitrosamides are direct acting (needing no metabolic activation) carcinogens and since high concentrations of peptides are likely to be present in the human stomach (especially after a protein-rich meal), the possibility of in vivo formation of nitrosamides in the human stomach should be a matter of concern.[32] Chow et al. and Challis et al. have presented strong evidence for the formation of nitrosopeptides in model system studies.[30,31] The latter group of workers have also synthesized and studied the stability of *N*-(*N*-acetylprolyl)-*N*-nitrosoglycine, believed to be the first authentic *N*-nitroso peptide ever synthesized.[30] However, no report of the occurrence of these compounds in foods or their formation in the human stomach has been published.

Outram and Pollock reported that nitrosation of peptides under gastric conditions led to the formation of *N*-nitrosoiminodialkanoic acids which are quite different from nitrosamides.[33] These researchers postulated the mechanism as shown in Figure 2 for the formation of *N*-nitrosoiminodialkanoic acids. On the other hand, Kubacki et al. noted that when di- or tripeptides containing proline as an N-terminal amino acid were nitrosated, only the $-NH$ group in the N-terminal proline was nitrosated; the peptide $-NH$ group was left intact (not nitrosated).[34] It appears, therefore, that a great deal of uncertainty still exists, and further research is warranted in this important area.

E. Catalysis

Nitrosation of secondary amines is catalyzed by a number of nucleophilic anions such as SCN^-, I^-, Br^-, Cl^-, as well as by formaldehyde (an electrophile), thiourea, and certain nitrosophenols.[1] Since many of these anions or compounds are known to occur in foods or present in human saliva they could exert a significant influence on both the in vitro (during processing or storage of foods) and in vivo formation of nitrosamines[35,36] As pointed out earlier (Equation 2) the above-mentioned anions could react with H_2ONO^+ to form powerful nitrosating species having the general formula NOX; X^- being SCN^-, Cl^-, Br^-, etc. The order in which these anions are effective as catalysts is represented as follows:[1,9,35-37]

$$SO_4^{-2} < Cl^- < Br^- < I^- < SCN^-$$

Since the formation of NOX is favored by lowering pH (due to the presence of more H_2ONO^+) these catalysts are highly effective below pH 3. Or, in other words, the pH optimum of nitrosation in the presence of the catalyst is shifted to a lower pH. Thus, the pH optima for nitrosation of N-methylaniline and morpholine in the presence of SCN^- were found to shift to pH 1.0 and 2.3, respectively, instead of the normal (uncatalyzed) pH optimum of 3 to 3.4.[36,37] At pH 1.5, 1 mM SCN^- increased the rate of nitrosation of N-methylaniline by 550-fold.[37] In contrast to the uncatalyzed reaction (Equation 6) the nitrosation rate in the presence of these catalysts is first order with respect to the concentration of each reactant, namely, the amine, nitrite, and the catalyst.[7] The secondary amides and the ureas, on the other hand, do not exhibit any anion-induced catalysis.[1]

Since the concentration of SCN^- in the saliva of smokers is about four to five times higher than that in nonsmokers a greater possibility of in vivo formation of nitrosamines would exist in the stomach of the former than the latter group of people.[37] Recent studies by Ladd et al. and Brunnemann et al. with human volunteers confirmed this hypothesis.[38,39] These researchers observed a significant increase in urinary excretion of nitrosoproline (NPRO) in smokers than nonsmokers after an oral dose of proline and sodium nitrate. As will be discussed later, urinary NPRO level can be used as an index of in vivo N-nitrosation occurring in the stomach due to the reaction of proline with salivary nitrite (formed from ingested sodium nitrate). Ladd et al. also noted a high degree of correlation of SCN^- levels in the saliva of the volunteers with urinary NPRO levels.[38]

Thiourea has been reported to be a stronger catalyst of N-nitrosation than thiocyanate.[40] It is effective even under mild acidic pHs (e.g., pH 4) which are more relevant to the conditions existing with foods. The greater catalytic effect of thiourea compared with that of SCN^-, Br^-, etc. is believed to be due to the large equilibrium constant for the formation of the sulfonium ion, $(NH_2)_2 \overset{+}{C} = S$ which has been suggested as the active nitrosating agent.[40]

The nitrosation of secondary amines can also be catalyzed by electrophilic species such as formaldehyde and chloral.[41] The initial formation of a positively charged iminium ion (Figure 3) is believed to help overcome the electronegativity of the amine nitrogen thus making it more susceptible to attack by the negatively charged nitrite ion which is otherwise inactive as a nitrosating agent in the absence of acid (H^+). It can be seen from the scheme in Figure 3 that H^+ is not involved in the reaction, and, therefore, the reaction can proceed even at neutral or alkaline pHs. This has actually been confirmed to be true in experiments. Acetaldehyde and other carbonyl compounds such as acetone are devoid of such catalytic activities. Since many smoked foods and fish products contain formaldehyde it might be involved in the catalysis of nitrosation during processing and storage of these food products.[1,41]

Certain phenols have been reported to catalyze or inhibit nitrosation depending on their relative concentration with respect to nitrite.[42-45] When excess phenols are present most of the nitrite is used up for C-nitrosation of the phenols or formation of quinone thereby inhibiting nitrosamine formation rather than catalyzing the reaction. On the other hand, when the concentration of phenol is very low a strong catalytic effect is observed. The nitrosophenol formed is believed to be the actual catalytic agent, and the reaction proceeds most likely according to the mechanism shown in Figure 4.[43] Since both phenols and nitrosophenols have been shown to be present in cured smoked meats and since various phenolic compounds (e.g., kaempferol, quercitin, catechin) occur naturally in many vegetables and plants, their presence in foods could have a significant effect on both the in vitro and in vivo formation of nitrosamines.[42-46]

Okun and Archer showed that certain cationic surfactants such as decyltrimethylammonium bromide, phosphatidylcholine, Triton X®-100 can catalyze nitrosation of secondary amines, especially that of long chain dialkylamines.[47] The rate of nitrosation of dihexylamine at pH 3.5 was shown to increase approximately 800-fold in the presence of decyltrimethylammonium bromide. Lecithin, a constituent of many foodstuffs, also enhanced nitrosation of

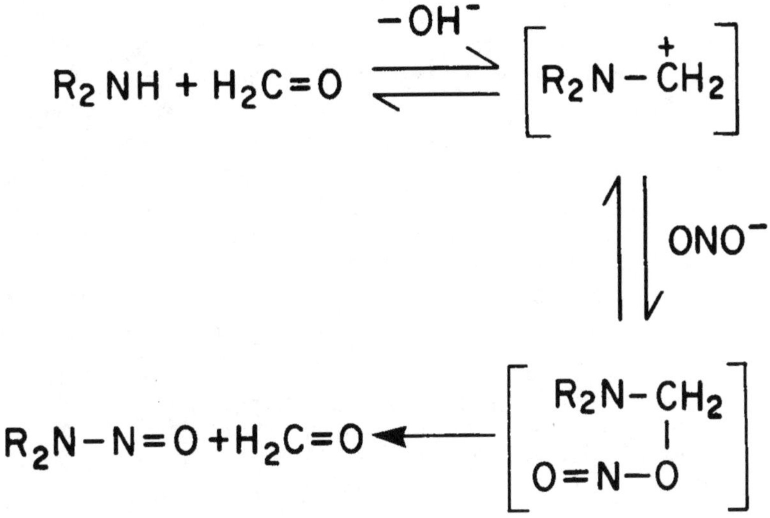

$$R_2NH + H_2C=O \; \underset{\xrightarrow{\hspace{1cm}}}{\overset{-OH^-}{\rightleftharpoons}} \; \left[R_2N - \overset{+}{C}H_2 \right]$$

$$\Big\Updownarrow \; ONO^-$$

$$R_2N-N=O + H_2C=O \longleftarrow \left[\begin{array}{c} R_2N-CH_2 \\ | \\ O=N-O \end{array} \right]$$

FIGURE 3. Proposed mechanism of the catalytic effect of formaldehyde on the nitrosation of a secondary amine. (From Roller, P. P. and Keefer, L. K., Catalysis of Nitrosation Reactions by Electrophilic Species, Publ. No. 9, International Agency for Research on Cancer, Lyon, 1974, 86.)

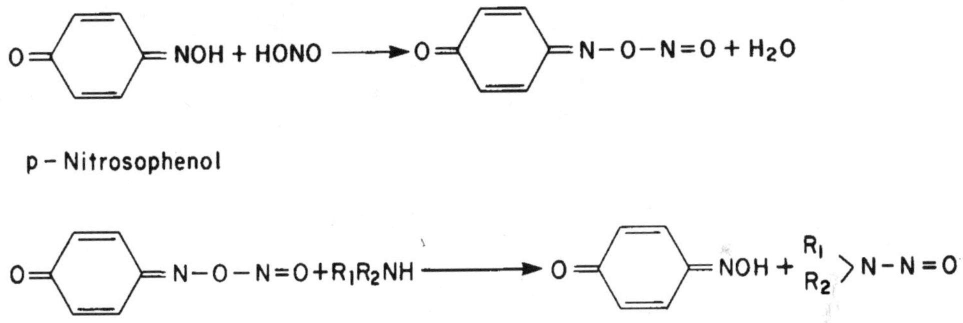

$$O= \!\!\bigcirc\!\! =NOH + HONO \longrightarrow O= \!\!\bigcirc\!\! =N-O-N=O + H_2O$$

p – Nitrosophenol

$$O= \!\!\bigcirc\!\! =N-O-N=O + R_1R_2NH \longrightarrow O= \!\!\bigcirc\!\! =NOH + \begin{array}{c} R_1 \\ R_2 \end{array}\!\!>\!N-N=O$$

FIGURE 4. Catalytic effect of *p*-nitrosophenol on *N*-nitrosation of secondary amines. (From Pignatelli, B., Friesen, M., and Walker, E. A., The Role of Phenols in Catalysis of Nitrosamine Formation, Publ. No. 31, International Agency for Research on Cancer, Lyon, 1980, 95.)

dihexylamine. Such an enhancement of N-nitrosation by various surfactants was explained in terms of electrostatic and hydrophobic interactions between the micellar phase and the reactants.

Considerable controversy exists in the literature as to the true nature (enzyme or nonenzymic) of bacteria-mediated catalysis of nitrosation of secondary amines. The subject has been recently reviewed by Archer.[10] As pointed out by Collin-Thompson et al. and Scanlan, various bacteria isolated from meat products could effect nitrosation of secondary amines by (1) reducing nitrate to nitrite, (2) synthesizing metabolites that may act as nonenzymic catalysts, or (3) favor nitrosation by decreasing the pH of the medium.[28,48] While studying nitrosation of dihexylamine Yang et al. also reached a similar conclusion.[49] The catalytic effect was found to be mainly nonenzymic in nature and involved hydrophobic interactions of the precursor amines and cellular constituents. Recent studies by Suzuki and Mitsuoka, however, indicate that certain intestinal bacteria are capable of catalyzing nitrosamine formation by a true enzymatic pathway.[50] The enzyme has a pH optimum of 7.4 and is destroyed by boiling in water. Thus, these bacteria may play a role in the in vivo formation of nitrosamines.

F. Inhibition

Any reaction or process that destroys a nitrosating agent or makes an amine precursor unreactive towards nitrosation (e.g., by lowering the pH and protonating the amine) will have a significant inhibitory effect on nitrosamine formation. Of all the known inhibitors ascorbic acid has been investigated most extensively followed by α-tocopherol. Both these compounds react extremely rapidly with nitrous acid (or other nitrosating agents) and convert it to nitric oxide which is inactive by itself as a nitrosating agent, but in the presence of oxygen nitric oxide reverts back to nitrogen dioxide and then to N_2O_3, a powerful nitrosating agent.[51-54] Therefore, to be effective as an efficient inhibitor an excess of these reductants must be present and oxygen should be excluded from the system. Ascorbic acid has an advantage over α-tocopherol in the sense that it is effective as an inhibitor at both acidic and alkaline pHs whereas the latter is only effective under acidic conditions. In the process of oxidation by nitrous acid ascorbic acid and α-tocopherol are converted to dehydroascrobate and α-tocoquinone, respectively. Although α-tocopherol is a phenolic compound it is incapable of forming a C-nitroso compound because all the positions in the aromatic ring are occupied by other groups.[1,52]

Thiols (e.g., cysteine, glutathione) can also inhibit *N*-nitrosation by reacting with nitrous acid to form thionitrite esters which, on the other hand, could also act as a trans-nitrosating agent especially at pH > 5. Like ascorbate and α-tocopherol, thiols, sulfur dioxide, bisulfite, 1,4-dihydroxyphenols reduce nitrous acid to nitric oxide.[1] Since many of these inhibitors are naturally occurring or used as food additives their presence in foods undoubtedly will influence formation of nitrosamines both in vitro and in vivo.

Other chemicals such as amino acids (containing a primary amino group), ammonia, and especially urea, hydrazine salts, hydrazoic acid, and sulfamic acid are also very effective in destroying nitrous acid. Ammonia, and primary amino acids are, however, poor inhibitors of nitrosation because they are protonated easily even under mildly acidic conditions, and, therefore, the reaction rate with nitrous acid is too slow. Most of the above compounds destroy nitrous acid by converting it to nitrogen thus making the destruction a more complete process. The last four groups of compounds, although extremely effective, have little use as an inhibitor for food use because of their toxic nature.

Ascorbate or its isomer erythorbate (also called isoascorbate) and α-tocopherol (vitamin E) are the two inhibitors which have been shown to give promising results in inhibiting nitrosamine formation in food systems, especially cured meats and bacon. Following the first reported observation by Mirvish et al. of the *N*-nitrosation inhibitory property of ascorbate Fiddler et al. demonstrated that formation of NDMA in frankfurters could be inhibited quite effectively by the addition of about 550 ppm sodium ascorbate or erythorbate.[51,54] Similar tests with bacon, however, were less fruitful.[53,55] The inability of ascorbate to effectively block NPYR formation in fried bacon was thought to be due to its limited solubility in fat. Since most of the NPYR in fried bacon has been shown to be formed in the fat phase (during frying) ascorbate is unable to reach the reaction sites, and, therefore, it is only partially effective.

Sen et al. working with bacon and Gray et al. working with model systems were first to demonstrate that other antioxidants such as ascorbyl palmitate, propyl gallate, and butylated hydroxyanisole, all of which are more fat soluble than ascorbate, were much more effective in inhibiting nitrosamine formation during frying of bacon.[55,56] Ascorbyl palmitate was, however, later found to have a major drawback in that it was unstable in bacon after storage, and, therefore, its inhibitory effect deteriorated with time. Other researchers later observed that α-tocopherol was much superior to other inhibitors mainly for three reasons.[56-58] It was highly fat soluble, it was found to be quite stable upon storage, and it reacted with nitrite only during frying of bacon. Thus it did not destroy the nitrite in bacon that was primarily added to prevent the outgrowth of botulinum spores. On the other hand, ascorbate was found to react easily with nitrite during storage of bacon, and, therefore, there was a danger of interference with the antibotulinal property of nitrite (if excess ascorbate were used). Gray et al. developed a technique by which α-tocopherol, at an equivalent concentration of 500 ppm in the final raw product, could be applied to both dry- and brine-cured bacon in the form of α-tocopherol-coated salt.[58] The products after frying were found to contain significantly lower levels of nitrosamines than comparable control samples prepared without α-tocopherol. Thus, at the moment α-tocopherol seems to be the most promising inhibitor of nitrosamine formation in bacon that could be used commercially and seems to be reasonably safe for food use.

The inhibitory effect of ascorbic acid and various other inhibitors on the in vivo formation of nitrosamines in experimental animals and man has been studied extensively. A useful summary of these findings has been compiled by Douglas et al.[9] Most of the in vivo results support those observed in in vitro experiments.

Ohshima et al. have reported an ingenious and novel approach for measuring in vivo nitrosation in man and laboratory animals.[59] They have shown that nitrosamino acids, mainly NPRO and *N*-nitrososarcosine (NSAR), are not usually metabolized and are excreted nearly quantitatively in the urine and feces (only minor amounts). When proline and nitrate (or vegetables containing nitrate) are administered orally to man or animals any NPRO detected in the urine can be taken as an indicator of the in vivo formation of NPRO (nitrite is produced in the salivary gland by reduction of the ingested nitrate). Simultaneous administration of ascorbic acid has been shown to significantly inhibit such in vivo formation of NPRO in man. Similarly, in vivo nitrosation of proline in rats has been shown to be catalyzed or inhibited (depending on the relative ratios of phenolics to nitrite) by several phenolic compounds as would have been predicted from the results of the in vitro studies (see earlier section).[60]

Stitch et al. also observed strong inhibitory effects of various dietary phenolic compounds (caffeic and ferulic acids, those present in tea) on in vivo formation of NPRO in man.[61] These workers did not observe any catalytic effect by the phenolic compounds tested because the amounts of various phenolics (as such or as present in tea) ingested always exceeded that of nitrite. These studies clearly demonstrate the important role of diet (e.g., tea, coffee, vegetables, citrus fruits, milk, and others containing naturally occurring nitrite scavengers) on the inhibition of in vivo *N*-nitrosation in man. It is tempting to conclude that nature has provided us with built-in protection against the hazard of in vivo formation of nitrosamines which might otherwise occur in significant amounts due to consumption of foods containing high levels of nitrate (e.g., green vegetables).

III. PRECURSORS OF NITROSAMINES IN FOODS

A detailed discussion of various precursors of *N*-nitroso compounds is beyond the scope of this article. The reader should consult some of the recent review papers for details.[1,28] Although considerable data are available as to the sources of nitrates and nitrites only a limited amount of data are available on the nitrosatable amines and amides in foods. Nitrates and nitrites in our diet could originate in various ways: as additives in cured meats and cheeses or come from natural sources such as that occurring in vegetables, drinking waters, etc. It has been estimated that approximately 86% of our total dietary intake of nitrates originates from vegetables and only 2% from cured meats. On the other hand, cured meat is the main source of the preformed nitrite because of its wide use as a preservative in these products. The permissible levels of nitrite in cured meats in various countries vary widely, but usually range between 100 to 200 ppm (as Na^+ or K^+ salts). Since nitrite is highly reactive, its concentration in cured meats gradually decreases with storage and depends on many factors such as processing conditions, duration and condition of storage, and cooking methods. Reliable data on the actual levels of nitrate and nitrite in cured meats as consumed by an average consumer (i.e., concentration just prior to eating) are, however, lacking. Further data in this areas will be highly desirable.

Vegetables usually do not contain significant levels of nitrite but traces have been reported to occur in temperature-abused carrot and spinach. Per capita daily intake of nitrite by an average U.S. resident has been estimated to be about 0.77 mg of which approximately 39% originates from cured meats and 34% from baked goods and cereals.[1] These data only relate to exogenous sources; additional amounts are ingested as salivary nitrite resulting from reduction of nitrate (absorbed from the GI tract) in the salivary glands. Thus, if one considers the total body burden of nitrite the major (3.5 mg out of a total of 4.2 mg) portion seems to originate from salivary nitrite of which 3 mg is of vegetable origin. This, however does not necessarily mean that all this nitrite (salivary nitrite of vegetable origin) will be available for in vivo formation of nitrosamines. As pointed out earlier, because of the concomitant presence of vitamin C and many phenolic compounds in these vegetables this nitrite might be rapidly destroyed in the stomach before it has a chance to form *N*-nitroso compounds.

Apart from data on the occurrences of dimethylamine, trimethylamine, and trimethylamine oxide in various fish and seafoods, extensive data on the occurrence of nitrosatable amines in other foods or beverages are scarce. Some limited data on secondary amine levels in cheeses, meats, vegetables, coffee, and that of proline and hydroxyproline in pork bellies are, however, available. These findings have been reviewed recently.[14,15,67]

Very little is known on the occurrence of alkylureas or other nitrosatable amides in foods. Kawabata et al. have failed to detect any alkylureas in 57 samples of various foods analyzed.[62] Methylguanidine, which has been reported to occur in meats and seafoods, may be nitrosated to form methylnitrosocyanamide and *N*-nitrosomethylurea (NMU).[63] Similarly, creatinine, which occur in high concentrations in cooked or stored meat and fish products, can be nitrosated in the stomach to form NSAR and NMU.[64,65] However, the yields of formation of both these nitroso compounds have been found to be extremely low.

Weisburger et al. have presented evidence for the formation of alkylnitrosoureido type compounds after nitrosation of certain dried fish extracts, but the exact nature of the nitrosourea(s) or its precursor was not identified.[66] Other amide precursors such as N-substituted amides could be formed in foods during high heat cooking (e.g., roasting, frying) as a result of reaction to α-amino acids with fatty acid esters.[67] It is highly unlikely, however, that significant amounts of such amides would be formed under normal cooking conditions.

IV. OCCURRENCES OF NITROSAMINES IN FOODS

The first reported occurrence of extremely high levels of NDMA in nitrite-preserved herring meal, which was implicated in the outbreak of liver disease and death of fur-bearing animals in Norway around the early 1960s, is now history.[68] This discovery led to a frantic search throughout the world for the possible presence of NDMA and other nitrosamines in human foods, especially those preserved with nitrite such as cured meats and fish. Studies carried out during the past 20 years have revealed the presence of various nitrosamines in a wide variety of foods and beverages. The reported findings are too numerous to mention individually. The details can be obtained from some of the earlier reviews and the original literature quoted under references.[28,67,69]

A. Nitrosamines in Cured Meats and Fried Bacon

Table 2 gives the wide ranges of different *N*-nitroso compounds which have been reported to occur in these products. Since the levels of these compounds in both cured meats and fried bacon have decreased significantly during the last 10 years the ranges for the levels for various *N*-nitroso compounds, as indicated in Table 2, may not represent the situation existing today; the last column gives that information. The general decline in the levels of nitrosamines in these products can be attributed to various factors such as better control of the input of nitrite by the industry, a decrease in the permissible levels of nitrate and nitrite, a ban on the use of nitrite-spice premixes, and the increased use of ascorbates, etc.[70]

As can be seen from the data in Table 2, fried bacon and some dry-cured meat products (ham and bacon) still contain significant levels of nitrosamines. Continuous monitoring of nitrosamine levels in fried bacon in both Canada and the U.S. during the last 10 years have indicated a significant decline in the levels of nitrosamines in this product.[71,72] However, high levels are still detected occasionally in some samples. The problem of high nitrosamine levels in dry-cured hams and bacons after frying can be best exemplified from the study by the U.S. Nitrite-Safety Council that reported the presence of alarmingly high levels of nitrosamines (up to 320 ppb) in some samples.[73] The exact reason for this is not clear. Localized concentration of high levels of nitrite may be one explanation (because of inhomogeneous nature of the dry curing technique). Such products are rare in Canada. The present average levels of various nitrosamines in the other products (Table 2) appear to be negligible.

Most of the nitrosamines detected in cured meats and bacon are powerful carcinogens with the exception of NSAR, which is a weak carcinogen, and NPRO which is noncarcinogenic.[4] As mentioned earlier, no toxicity data on *N*-nitrosothiazolidine (NThZ), *N*-nitrosothiazolidine-4-carboxylic acid (NTCA), and *N*-nitrosohydroxyproline (NHPRO) are available. The data on the occurrences of NThZ and NTCA are also very limited because both these compounds have been discovered (in these products) only very recently (see Addendum).

Recent data from the University of British Columbia suggest that certain cured meat products (canned meat, Chinese sausage) may contain high levels (up to 3900 ppb) of NPRO.[74] A major portion of this NPRO in some samples was found to be present in bound form (e.g., as nitrosated peptides) that was not measurable by the normal method. Although NPRO is noncarcinogenic it can lead to the formation of carcinogenic NPYR upon cooking at a high temperature.[71,75] Therefore, the presence of such high levels of NPRO in cured meats should be viewed with concern.

B. Mechanism of Formation of NPYR in Fried Bacon

Of all the cured meat products, fried bacon is the only item which has been found to contain nitrosamines, mainly NPYR and NDMA, quite consistently. Usually, the raw bacon is free of nitrosamines; they are formed only during frying. Various factors such as the concentration of nitrite (both initial and residual levels), processing conditions, diet of the pigs (see later), the lean to adipose tissue ratio, the presence of inhibitors, frying temperature, and cooking methods control the formation of nitrosamines.[67] In general , a fatty bacon processed with high levels of nitrite and fried in a conventional frying pan at $\approx 170°C$ to a well-done stage contain the highest levels of nitrosamines. Bacons cooked in a microwave oven contain significantly lower levels of nitrosamines. Since most of the nitrosamines detected in fried bacon are steam-volatile the major portions are volatalized in the fumes produced during frying. Also, the cooked-out fat contains higher (approximately twice) levels of nitrosamines than the cooked lean.

Considerable research has been carried out to determine the mechanism of formation of NPYR in bacon. The subject matter has been recently reviewed by Gray.[67] Among the many theories proposed, that involving nitrosation of proline followed by decarboxylation of NPRO to NPYR during heating seems to be most popular. However, failure to detect significant levels of preformed NPRO in raw bacon and an observed low conversion rate (~0.16%) of NPRO to NPYR during frying of bacon seems to rule out preformed NPRO being the main precursor. As Gray has pointed out additional amounts of NPRO are probably formed during frying that might account for the high levels of NPYR detected in fried bacon.[67] Other amine precursors proposed include (1) putrescine, (2) proline-containing peptides, and (3) pyrrolidine. All of these precursors have been shown to form NPYR upon heating with nitrite.[28]

The nitrosating agent responsible for the formation of nitrosamines in fried bacon might be N_2O_3, formed during heating of nitrite in bacon, or nitric oxide radical (NO) formed by

Table 2
REPORTED OCCURRENCES OF *N*-NITROSO COMPOUNDS IN CURED MEATS AND BACON[69,73-82]

Type	*N*-Nitroso[a] compounds detected	Highest level reported	Levels (ppb) Range	Approx. average level at the present time
Sausages (smoked and unsmoked dry and semi-dry), bologna, meat loaves, salami, and European-type sausages (some heavily spiced)	NDMA. NPYR. NPIP. NMOR	300 ppb NPIP in Thuringer liver sausage. 84 ppb NDMA in frankfurter. 105 ppb NPYR in Mettwurst sausage	Tr.[b]—300	Tr.
	NHPYR. NHPRO. NPRO. NSAR	401 ppb NPRO in bologna	Tr.—401	Tr.
Fried bacon	NDMA. NTCA NThZ. NTCA	944 ppb NTCA in smoked sausage	Tr.—944	?
	NDMA. NDEA. NPIP. NPYR NMOR	108 ppb NPYR 39 ppb NDMA	Tr.—108	10 ppb NPYR
Raw bacon	NHPYR. NPRO. NThZ. NTCA	32 ppb NThZ 67 ppb NPRO	Tr.—67	?
Cooked-out bacon fat	NPRO. NThZ NDMA. NDEA. NPIP. NPYR NHPYR. NThZ	44 ppb NPRO. 11 ppb NThZ 207 ppb NPYR 30 ppb DNMA	Tr.—44 Tr.—207	? 20 ppb NPYR
Hams (smoked and unsmoked; some spiced), dry-cured bacon and ham (both cooked and uncooked)	NDMA. NPIP. NPYR	NPYR 320 ppb in dry-cured bacon, 64 ppb NPIP in spiced ham	Tr.—320	~10 ppb in dry-cured products
Miscellaneous cured meats (picnics, corned beef, ham paste)	NHPYR. NPRO. NThZ. NTCA	604 ppb NPRO in dry-cured ham 617 ppb NPRO in smoked ham	Tr. 617	?
	NDMA. NPYR. NPIP	149 ppb NPYR in fried pork side meat 63 ppb NDMA in souse	Tr.—149	Tr.
	NPRO. NThZ NTCA	411 ppb NPRO in raw pork side meat 4400 ppb NTCA in smoked ham paste	Tr.—4400	?

Table 2 (continued)
REPORTED OCCURRENCES OF *N*-NITROSO COMPOUNDS IN CURED MEATS AND BACON[69,73-82]

Type	*N*-Nitroso[a] compounds detected	Levels (ppb)		Approx. average level at the present time
		Highest level reported	Range	
Canned meats (shelf-stable and pasteurized)			Mostly negative	
Baby foods containing meat products			Mostly negative	

[a] Abbreviations: *N*-nitrosopiperidine (NPIP), *N*-nitrosomorpholine (NMOR), *N*-nitrosohydroxypyrrolidine (NHPYR).
[b] Tr., traces (≤1 ppb).
[c] Not enough data available.

dissociation of N_2O_3 at high temperature ($>100°C$).[67] Recently, Dennis et al. reported a 50 to 90% reduction in the formation of both NDMA and NPYR if bacon was fried in a nitrogen atmosphere than in the presence of air (oxygen) as is normally done.[83] These authors concluded that the nitrosamines in bacon were formed by an oxygen-dependent mechanism, the key step being the oxidation of NO and the formation of higher nitrogen oxides which could act as direct nitrosating agents. Since the nitrosation by nitrogen oxides in the lipid phase is a very rapid process even a low level of pyrrolidine, present or formed by decarboxylation of proline during frying, could account for all the NPYR detected in fried bacon. It has been suggested that pseudonitrosites, formed by the interaction of unsaturated triglycerides and nitrite, could also act as nitrosating agents.[84] This might explain the reason for the observed formation of higher levels of nitrosamines in bacon from pigs fed a corn oil- (more unsaturated triglycerides) supplemented diet than that from pigs fed a coconut fat- (mostly saturated triglycerides) supplemented diet.[85]

Several researchers observed that no significant formation of NPYR took place during frying of bacon not until most of the moisture had been driven off from the system and the temperature of the fat phase had reached 150°C or higher.[66,86] In addition, Mottram et al. noted only negligible formation of NPYR if minced lean (from raw bacon) was heated in corn oil at 175°C because the presence of moisture prevented a rapid rise in temperature.[86] However, when freeze-dried lean was heated as above the temperature quickly rose resulting in the formation of both NDMA and NPYR. Such an observation is consistent with the nitrogen oxide theory as mentioned above. The above researchers concluded that the precursors for nitrosamine formation were present in both the fat and lean but the lipid phase provided an environment that was more conducive to nitrosamine formation. The presence of moisture in the lean probably destroyed the nitrogen oxides due to hydrolysis.

As can be seen from above, the formation of NPYR in bacon is very complex. Probably, a combination of several mechanisms is in operation, and a particular mechanism may predominate under a particular circumstance.

C. Nitrosamines in Foods Dried with Hot Flue Gases

Since hot flue gases may contain very high levels of NO_x, it is not surprising that many foods processed by the direct drying technique (using hot flue gases) have been reported to contain significant levels of nitrosamines.[87,88] The possibility of such formation of nitrosamines was first reported by the Norweigian workers while investigating formation of NDMA in nitrite-treated herring meal and later by Sen et al. in connection with their work on the formation of NDMA in nonnitrited fish meal.[88] The problem received much more publicity when Spiegelhalder et al. discovered that malt dried by such a technique contained high levels (up to 600 ppb) of NDMA.[87] Beer and ale prepared from such malts were found to contain on the average 2.7 ppb NDMA, whereas beers produced from malt which had been dried using an indirect drying technique were either negative or contained extremely low levels of NDMA. It was postulated that NO_x gases (mostly NO_2) first reacted with the moisture on the wet malt to form nitrous acid which then transformed to N_2O_3. These workers also demonstrated that such formation of NDMA in the malt could be prevented or significantly reduced by burning extra sulfur with the fuel. Sulfur dioxide generated from burning the sulfur apparently acted as an inhibitor of nitrosation by reacting with the NO_x gases (see under Section II.F.

The levels of nitrosamines detected in malt and malt-based beverages (beer, ale, whisky, etc.) in different countries are presented in Table 3. It should be noted that due to the improvement in the drying techniques these levels have decreased significantly during the last 4 to 5 years.[89] The present average levels of NDMA in various beers and ales in Canada, U.S., Germany, and other countries are reported to be extremely low (≈0.1 ppb).[89] Table 3 also gives some data on the nitrosamine levels in a few other foods (e.g., skim milk

Table 3
OCCURRENCES OF NITROSAMINES IN MALT, MALT-BASED BEVERAGES, AND OTHER DIRECT DRIED FOODS

Item	Period of analysis	Nitrosamines detected	Level (ppb) Mean (estimated)	Level (ppb) Range
Malt[22,76,87,90]	1981 and earlier	NDMA	20	Tr.[a]—600
	1982—1983	NDMA	3	Tr.—9
	1983	NPRO	24	5—42
	1983	NSAR	?[b]	Tr.
Beer and ale[87,91,94]	Prior to 1982	NDMA	~2	Tr.—68
		NPYR	?	Tr.
	1982—1983	NDMA	<0.5	0—1.9
	1982—1983	NPRO	1.7	Tr.—6
Whisky[94]	1979	NDMA	Tr.	0—2
Skim milk powder[22,95-97]	1979-1981	NDMA	0.5	0—4.5
Instant and roasted coffee[22]	1980-1981	NDMA	?	Tr.
		NPYR	0.3	Tr.1.4
Dried foods containing soy protein[81]	1980—1981	NDMA	?	Tr.

[a] Tr., traces (≤ 1 ppb).
[b] ?. Not enough data available.

powder, dried soup bases, instant coffee, soy protein isolate) which are also processed sometimes using direct drying methods. Although these products contain only traces of nitrosamines and may not pose any significant health hazard to man, it might be advisable to reduce the concentration of nitrosamines in these products if technologically and economically feasible.

D. Mechanism of Formation of NDMA in Dried Malt

Although the nitrosating agent in malt has been clearly identified to be the NO_x gases the same thing can not be said about the amine precursors. Dimethylamine, hordenine, gramine, all of which are known to be present in germinated barley, have been proposed as the possible precursors.[87,98-99] Recent studies by Mangino et al. suggest that although hordenine is present in malt in higher concentrations than gramine, the former is unlikely to be the major precursor of NDMA because its nitrosation rate is too slow.[99] In aqueous solution, the nitrosatin rate of gramine (to form NDMA) is 5200 times faster than that of hordenine and is comparable to that of dimethylamine, a secondary amine. Based on these findings, these workers suggested that gramine and dimethylamine are the likely precursors of NDMA in malt.

From a detailed study of the kinetics of nitrosation and of the reaction products Mangino et al. concluded that gramine did not undergo nitrosation by the nitrosative dealkylation route, the common pathway for the nitrosation of tertiary amines. To explain the facile nitrosation of gramine they proposed a new mechanism as illustrated in Figure 5. Since the mechanism is based on nitrosation in aqueous acid it is not certain whether the same mechanism would apply to conditions existing in malt drying operations.[99]

E. Nitrosamines in Other Foods

Among the other foods, cheese and fish (both fresh and salted) have been investigated most extensively and occasionally found to contain traces of nitrosamines, mainly NDMA. But, except in sporadic instances, the levels of nitrosamines detected in both these products are quite low (≤ 1 ppb). Gray et al. and Sen have reviewed the literature dealing with the

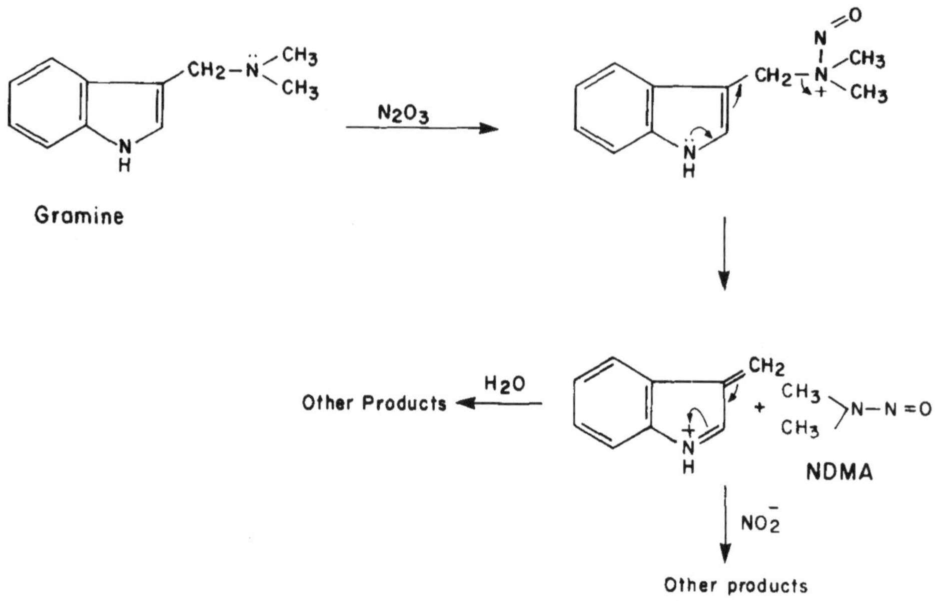

FIGURE 5.   Proposed mechanism for the nitrosation of gramine to yield NDMA. (From Mangino, M. M. and Scanlan, R. A., Rapid formation of NDMA from The Alkaloid Gramine: A Naturally Occurring Tertiary Amine Precursor of NDMA in Malt, Publ. No. 57, International Agency for Research on Cancer, Lyon, 1984, 337.)

occurrence and formation of nitrosamines in cheese and fish, respectively.[69,100] In addition, certain fish products, especially cod, herring, whiting, and crab have been found to form significant amounts of NDMA and NPRO when incubated under simulated gastric conditions.[101] Some samples yielded as much as 44 μg NDMA per portion (equivalent to an average meal). The researchers concluded that in vivo nitrosation of fish amines would pose a greater danger to human health than the low levels of preformed nitrosamines present in the sample.

Kawabata et al. observed that when certain seafoods (dried squid, mackerel, salt dried "shishyamo" and "sanmahiraki") were cooked (broiled) on a city gas range significant amounts (up to 313 ppb) of NDMA were formed during the cooking process.[23] The highest level of NDMA was found to be formed when kerosene was used as a fuel followed by that using a city gas range. This was explained by the finding that flue gases from a kerosene stove contained higher levels of NO_x than that from a gas stove. The principal amine precursor was believed to be trimethylamine oxide which is known to be present in these foods in high concentrations. Kawabata et al. also detected traces (<0.5 ppb) of NDMA and slightly higher levels (0.5 to 5 ppb) of NPYR in many traditional salt-fermented vegetables commonly consumed by Japanese people.[62] Since both the dried seafoods and pickled vegetables are consumed in large quantities in Japan the above findings may have important signifance to the health of the Japanese people.

The finding of exceedingly high levels of nitrosamines (NDMA, NPYR, and NPIP) in spice-nitrite premixes, which were widely used until 1974 in both Canada and U.S. for the preparation of a wide variety of cured meat products, has been discussed in previously published reviews.[28,69] Banning of these premixes has indeed removed a major source of nitrosamines in cured meats. Traces of NPIP which are still occasionally found in spiced meat products might be originating due to the interaction of amines in the spices and nitrite. However, both the incidence and the average level has been found to be quite low.

Two other instances of occurrence of nitrosamines in foods might be worth noting. In the first case, the finding of an unusual nitrosamine, *N*-3-methylbutyl-*N*-1-methylacetonylnitro-

Table 4

**AVERAGE PER CAPITA DAILY INTAKE OF VOLATILE
NITROSAMINES THROUGH FOOD AND BEVERAGES (EXOGENOUS
SOURCES ONLY)**

Country	Contribution from	μg Nitrosamine[a,b] ingested		Percent of total intake	
		For beer drinkers	For others	For beer drinkers	For others
Federal Republic of Germany[87,89]	Beer[b]	0.7	0	64	0
	Cured meats	0.21	0.21	10	47
	Cheese	0.01	0.01	0	2
	Other foods and beverages	0.23	0.23	25	51
The Netherlands[105]	Beer	0.99	0	90	0
	Other foods	0.11	0.11	10	100
U.K.[106]	Cured meats	0.43	0.43	—	81
	Beer	Data not available			
	Fish	<0.01	<0.01	—	<2
	Cheese	<0.01	<0.01	—	<2
	All other foods	0.08	0.08	—	15
Japan[104]	Beer	0.05	0	2	0
	Broiled dried and all other fish and seafoods	2.37[c]	2.37[c]	79	81
	Dairy (eggs, chicken)	0.53	0.53	18	18
	Others (including cured meats)	0.03	0.03	1	1

[a] For simplicity all volatile nitrosamines have been grouped together and the amounts have been expressed in terms of NDMA, the predominant nitrosamine detected.

[b] The 1983 figures for Germany indicated a total NDMA intake of 0.5 to 0.6 μg/person/day (instead of 1.15 μg as above) due to the significant reduction in the level of NDMA in beer.[89]

[c] Recent data by Yamamoto et al. suggest a daily per capita consumption of only 0.5 μg NDMA through consumption of fish and seafoods.[119]

samine, in nitrosated moldy corn is of great interest because the consumption of bread prepared from such moldy corn has been implicated in the high incidence of esophageal cancer among the inhabitants of Northern China, although no definite conclusion has yet been reached.[102] In the second case, the source of traces of NMOR in margarine and butter has been traced to wax coatings on the packaging papers used to wrap the products.[103] The observation has also been confirmed in the author's laboratory.[120]

F. Nitrosamines in Total Diet Samples

Thus far, only the occurrence of nitrosamines in specific food items has been discussed. Since food habits vary widely from person to person and among persons in different countries it is extremely difficult to obtain an accurate estimate of average per capita daily consumption of nitrosamines ingested through foods. Nevertheless, scientists in different countries have attempted to do so either by analyzing composite total diet samples or by estimating intake of nitrosamines using the average per capita consumption data for different commodities.[89,104-106] These data for the Federal Republic of Germany, The Netherlands, Japan, and the U.K. are given in Table 4. As expected, they do differ, although the figures for the three European countries are somewhat similar. The U.K. data did not include beer because the problem of NDMA contamination in this product was unknown at that time. For persons drinking beer regularly it is the main exogenous dietary source of nitrosamines. However, in many countries (e.g., those in Asia) the consumption of beer and cured meats would be negligible, and, therefore, the average daily intake of nitrosamines would be quite different. It should also be noted that the figures in Table 4 do not include nonvolatile nitrosamines, bound nitrosamines, nitrosamides, or the newly discovered NThZ and NTCA. The last two compounds might be present in smoked meats and fish in high concentrations. Moreover, many foods, especially cured meats and salted Japanese vegetables, contain high levels of unidentified Thermal Energy Analyzer-positive compounds which might be *N*-nitroso compounds.[107] It is apparent, therefore, that a lot of gaps still exist in our knowledge in this area. The actual amounts of total nitrosamines digested through foods might be much more than those suggested from the data in Table 4.

V. IN VIVO FORMATION OF NITROSAMINES IN MAN AFTER CONSUMPTION OF A MEAL

Various in vivo studies with laboratory animals have amply demonstrated the likelihood of formation of *N*-nitroso compounds in the human stomach from the ingested precursors present in our food supplies. Since the results of these studies have been reviewed elsewhere, they will not be discussed in this article.[1,4,7,9] Instead, the discussion will be restricted only to those carried out with human volunteers. As mentioned earlier, the most convincing proof of the in vivo formation of *N*-nitroso compounds in man came from the work of Ohshima and Bartsch, who demonstrated increased excretion of NPRO in the urine of a volunteer after oral administration of proline and nitrate.[59] Other workers later have confirmed the finding.[108,109] An increased excretion of NSAR after meals has also been observed in some cases but the source of the NSAR (if already present in the food) or of the amine precursor, sarcosine, has not been identified. Most recent work by Tsuda et al. and Ohshima and co-workers suggests that in addition to NPRO, NTCA and NMTCA are also formed in vivo in man in a similar manner after consumption of normal meals.[16,109] Thus, the above technique will be particularly useful for studying endogenous nitrosation in human subjects at high risk. Several interesting studies have already been carried out to investigate such possibilities.[110-111]

Although the above technique gives a good indication of in vivo *N*-nitrosation in general, it does not provide any direct evidence for the formation of the carcinogenic nitrosamines (e.g., NDMA, NPYR) in the stomach that are commonly found in foods. The main difficulty is that, in contrast to the nitrosamino acids, these nitrosamino are metabolized rapidly by the liver and, therefore, very little is detected in the blood or excreted in the urine. Although several researchers reported an increase in the concentration of NDMA and NDEA in the blood of human volunteers after consumption of meals consisting of beer, nitrate-rich vegetables, and cured meats (fried bacon, smoked meat), a great deal of controversy still exists as to the validity of the findings.[112-116] First of all, it is not certain whether the nitrosamines detected in the blood were originally present or formed as an artifact during the analysis. Using different methods of analysis other laboratories have been unable to detect any significant level of nitrosamines in similar samples.[116] Second, it has been suggested that consumption of ethanol (in beer) may have inhibited the metabolism of the nitrosamines by the liver thus making their detection in the blood possible.[117] But this does not explain the consistent finding of NDMA in the resting blood of many volunteers where the samples were collected before consumption of any meal or alcoholic beverages.[112,114,117] This is especially disturbing in view of the fact that some laboratories have failed to detect any NDMA or NDEA in the resting blood from many human volunteers.[113,116] Although the positive findings may be in doubt (due to the possibility of artifactual formation), the negative results should be reliable because no evidence for false-negative results has thus far been reported.

If the positive findings are true the following questions still remain to be answered. How is it possible for the nitrosamines to escape metabolism by the liver? Or is it possible that one is seeing only the tip of the iceberg, i.e., much more nitrosamines are formed in the stomach and only a small fraction, which escapes destruction by the liver, is being seen. It is also possible that the nitrosamines are formed at different sites (other than stomach) and entering the blood stream after the "first-pass clearance" by the liver. Until these questions are resolved it might be unwise to draw any conclusion from the results of these studies.

VI. CONCLUSIONS

It can be concluded from the above that research carried out during the past 10 to 15 years has provided valuable data on the levels of N-nitroso compounds in various foods and improved our knowledge and understandings regarding formation of these compounds. It is also gratifying to know that due to a better understanding of the mechanism of formation of these compounds it has been possible to take adequate measures that significantly reduced the concentration of these compounds in various foods and beverages. Although it might be debatable whether such low levels of various N-nitroso compounds in foods or that formed in vivo in the human stomach pose (or posed in the past when higher levels of nitrosamines were present in foods) a significant health hazard to man, no one would question the wisdom of reducing the concentration of these carcinogens in our food supplies. Additional efforts would be necessary to further reduce the concentration of various N-nitroso compounds, including NThZ and NTCA, in fried bacon, dry-cured meat products, and smoked meats and fish. Also, a great deal of research remains to be done in some important areas especially on the chemistry and formation of nitrosamides (including nitrosopeptides), characterization of the many unidentified nonvolatile N-nitroso compounds which are presently being detected in foods, and occurrence of various amine and amide precursors in a wide variety of foodstuffs. It is hoped that various research projects now underway in different laboratories would provide the necessary data to fill the gaps in our knowledge that exist at the moment.

VII. ADDENDUM

Since the completion of the manuscript, considerable amounts of new information have been published on the mechanism of formation as well as on the occurrence of NThZ and NTCA in smoked meats and fish.[121-124] Studies by Hotchkiss et al.[125] also suggest that cooking of certain foods in cooked-out bacon fat may lead to the formation of high levels of volatile nitrosamines. The nitrosating agent in the rendered bacon fat seems to be associated with the lipid fraction.[126] The reader should consult the original papers for details.

REFERENCES

1. National Research Council/National Academy of Sciences, The Health Effects of Nitrate, Nitrite, and N-Nitroso Compounds, National Academy Press, Washington, D.C., 1981, chap. 4 to 8.
2. **Scanlan, R. A. and Tannenbaum, S. R.**, *N-Nitroso Compounds*, ACS Symp. series No. 174, American Chemical Society, Washington, D.C., 1981.
3. **Preussmann, R.**, Public Health Significance of Environmental N-Nitroso Compounds, IARC Sci. Publ., No. 45, International Agency for Research on Cancer, Lyon, 1983, 3.
4. **Magee, P. N., Montesano, R., and Preussmann, R.**, N-Nitroso compounds and related carcinogens, in *Chemical Carcinogens*, ACS Monograph 173, Searle, C. E., Ed., American Chemical Society, Washington, D. C., 1976, 491.
5. **Ridd, J. H.**, Nitrosation, diazotisation and deamination, *Q. Rev.*, 15, 418, 1961.

6. **Mirvish, S.,** Kinetics of dimethylamine nitrosation in relation to nitrosamine carcinogenesis, *J. Natl. Cancer Inst.*, 44, 633, 1970.

7. **Mirvish, S.,** Formation of N-nitroso compounds: chemistry, kinetics, and *in vivo* occurrence, *Toxicol. Appl. Pharmacol.*, 31, 325, 1975.

8. National Academy of Sciences/National Research Council, Toxicants Occurring Naturally in Foods, Publ. 1354, National Academy of Sciences, Washington, D.C., 1966.

9. **Douglass, M. L., Kabacoff, B. L., Anderson, G. A., and Cheng, M. C.,** The chemistry of nitrosamine formation, inhibition and destruction, *J. Soc. Cosmet. Chem.*, 29, 581, 1978.

10. **Archer, M. C.,** Catalysis and Inhibition of N-Nitrosation Reactions, IARC Sci. Publ., No. 57, International Agency for Research on Cancer, Lyon, 1984, 263.

11. **Smith, P. A. S. and Loeppky, R. N.,** Nitrosative cleavage of tertiary amines, *J. Am. Chem. Soc.*, 89, 1147, 1967.

12. **Fiddler, W., Pensabene, J. W., Doerr, R. C., and Wassermann, A. E.,** Formation of N-nitrosodimethylamine from naturally occurring quaternary ammonium compounds and tertiary amines, *Nature (London)*, 236, 307, 1972.

13. **Ohshima, H. and Kawabata, T.,** Mechanism of N-Nitrosodimethylamine Formation from Trimethylamine and Trimethylamine Oxide, IARC Sci. Publ., No. 19, International Agency for Research on Cancer, Lyon, 1978, 143.

14. **Maga, J. A.,** Amines in foods, *Crit. Rev. Food Sci. Nutr.*, 10, 373, 1978.

15. **Smith, T. A.,** Amines in foods, *Food Chem.*, 6, 169, 1980—81.

16. **Ohshima, H., O'Neill, I. K., Friesen, M., Pignatelli, B., and Bartsch, H.,** Presence in Human Urine of a New Sulfur-Containing N-Nitrosamino Acid; N-Nitrosothiazolidine-4-Carboxylic Acid and N-Nitroso-2-Methylthiazolidine-4-Carboxylic Acid, IARC Sci. Publ., No. 57, International Agency for Research on Cancer, Lyon, 1984, 77.

17. **Challis, B. C. and Kyrtopoulos, S. A.,** Nitrosation under alkaline conditions, *J. Chem. Soc. Chem. Commun.*, p. 877, 1976.

18. **Challis, B. C., Edwards, A., Huma, R. R., Kyrtopoulos, S. A., and Outram, J. R.,** Rapid Formation of N-Nitrosamines from Nitrogen Oxides under Neutral and Alkaline Conditions, IARC Sci. Publ., No. 19, International Agency for Research on Cancer, Lyon, 1978, 127.

19. **Challis, B. C. and Outram, J. R.,** The chemistry of Nitroso-compounds. XV. Formation of N-nitrosamines in solution from gaseous nitric oxide in the presence of iodine, *J. Chem. Soc. Perkin Trans. 1*, p. 2768, 1979.

20. **Sen, N. P., Schwinghamer, L. A., Donaldson, B. A., and Miles, W. F.,** N-Nitrosodimethylamine in fish meal, *J. Agric. Food Chem.*, 20, 1280, 1972.

21. **Spiegelhalder, B., Eisenbrand, G., and Preussmann, R.,** Occurrence of Volatile Nitrosamines in Food: A Survey of the West German Diet, IARC Sci. Publ., No. 31, International Agency for Research on Cancer, Lyon, 1980, 467.

22. **Sen, N. P. and Seaman, S.,** Volatile nitrosamines in dried foods, *J. Assoc. Off. Anal. Chem.*, 64, 1238, 1981.

23. **Kawabata, T., Matsui, M., Ishibashi, T., Hamano, M., and Ino, M.,** Formation of N-Nitroso Compounds during Cooking of Japanese Food, IARC Sci. Publ., No. 41, International Agency for Research on Cancer, Lyon, 1982, 287.

24. **Helgason, T., Ewen, S. W. B., Jaffray, B., Stowers, J. M., Outram, J. R., and Pollock, J. R. A.,** Nitrosamines in Smoked Meat and their Relation to Diabetes, IARC Sci. Publ., No. 57, International Agency for Research on Cancer, Lyon, 1984, 911.

25. **Hendrickson, J. B., Cram, D. J., and Hammond, G. S.,** *Organic Chemistry*, 3rd ed., McGraw-Hill, New York, 1970, 413 and 540.

26. **Warthesen, J., Scanlan, R. A., Bills, D. D., and Libbey, L. M.,** Formation of heterocyclic N-nitrosamines from the reaction of nitrite and selected primary diamines and amino acids, *J. Agric. Food Chem.*, 23, 898, 1975.

27. **Hildrum, K. I., Scanlan, R. A., and Libbey, L. M.,** Identification of γ-butenyl-(β-propenyl)-nitrosamine, the principal volatile nitrosamine formed in the nitrosation of spermidine or spermine, *J. Agric. Food Chem.*, 23, 34, 1975.

28. **Scanlan, R. A.,** N-Nitrosamines in foods, *CRC Crit. Rev. Food Technol.*, 5, 357, 1975.

29. **Pollock, J. R. A.,** Nitrosation Products from Peptides, IARC Sci. Publ., No. 41, International Agency for Research on Cancer, Lyon, 1982.

30. **Challis, B. C., Hopkins, A. H., and Milligan, J. R.,** Nitrosation of Peptides, IARC Sci. Publ., No. 57, International Agency for Research on Cancer, Lyon, 1984, 61.

31. **Chow, Y. L., Dhaliwa, S. S., and Polo, J.,** A Chemical Model of Nitrosamide Carcinogenesis: the Nitrosation of Amide Linkages and Facile Transformation of Nitrosamides, IARC Sci. Publ., No. 57, International Agency for Research on Cancer, Lyon, 1984, 317.

32. **Challis, B. C., Lomas, S. J., Rzepa, H. S., Bavin, P. M. G., Darkin, D. W., Viney, N. J., and Moore, P. J.**, A kinetic model for the formation of gastric N-nitroso compounds, in *Nitrosamines and Human Cancer, Banbury Report No. 12*, Magee, P. N., Ed., Cold Spring Harbor Laboratory, N. Y., 1982, 243—256.

33. **Outram, J. R. and Pollock, J. R. A.**, Production of N-nitrosoiminodialkanoic Acids by Nitrite in Gastric Juice, IARC Sci. Publ., No. 57, International Agency for Research on Cancer, Lyon, 1984, 71.

34. **Kubacki, S., Havery, D. C., and Fazio, T.**, Determination of Nonvolatile N-Nitrosamines in Foods, IARC Sci. Publ., No. 57, International Agency for Research on Cancer, Lyon, 1984, 145.

35. **Boyland, E. and Walker, S. A.**, Thiocyanate Catalysis of Nitrosamine Formation and Some Dietary Implications, IARC Sci. Publ., No. 9, International Agency for Research on Cancer, Lyon, 1974, 132.

36. **Fan, T. Y. and Tannenbaum, S. R.**, Factors influencing the rate of formation of nitrosomorpholine from morpholine and nitrite: acceleration by thiocyanate and other anions, *J. Agric. Food Chem.*, 21, 237, 1973.

37. **Boyland, E. and Walker, S. A.**, Effect of thiocyanate on nitrosation of amines, *Nature (London)*, 248, 601, 1974.

38. **Ladd, K. F., Newmark, H. L., and Archer, M. C.**, Increased Endogenous Nitrosation in Smokers, IARC Sci. Publ., No. 57, International Agency for Research on Cancer, Lyon, 1984, 811.

39. **Brunnemann, K. D., Scott, J. C., Haley, N. J., and Hoffmann, D.**, On the Endogenous Formation of N-Nitrosoproline upon Cigarette Smoke Inhalation, IARC Sci. Publ., No. 57, International Agency for Research on Cancer, Lyon, 1984, 819.

40. **Masui, M., Fujisawa, H., and Ohmori, H.**, N-Nitrosodimethylamine formation catalyzed by alkylthioureas: a kinetic study, *Chem. Pharm. Bull.*, 30, 593, 1982.

41. **Roller, P. P. and Keefer, L. K.**, Catalysis of Nitrosation Reactions by Electrophilic Species, IARC Sci. Publ., No. 9, International Agency for Research on Cancer, Lyon, 1974, 86.

42. **Davies, R. and McWeeny, D. J.**, Catalytic effect of nitrosophenols on N-nitrosamine formation, *Nature (London)*, 266, 657, 1977.

43. **Walker, E. A., Pignatelli, B., and Castegnaro, M.**, Catalytic effect of p-nitrosophenol on the nitrosation of diethylamine, *J. Agric. Food Chem.*, 27, 393, 1979.

44. **Pignatelli, B., Friesen, M., and Walker, E. A.**, The Role of Phenols in Catalysis of Nitrosamine Formation, IARC Sci. Publ., No. 31, International Agency for Research on Cancer, Lyon, 1980, 95.

45. **Pignatelli, B., Bereziat, J. C., O'Neill, I. K., and Bartsch, H.**, Catalytic Role of some Phenolic Substances in Endogenous Formation of N-Nitroso Compounds, IARC Sci. Publ., No. 41, International Agency for Research on Cancer, Lyon, 1982, 413.

46. **Knowles, M. K., Gilbert, J., and McWeeny, D. J.**, Phenols in smoked, cured meats: nitrosation of phenols in liquid smokes and in smoked bacon, *J. Sci. Food Agric.*, 26, 267, 1975.

47. **Okun, J. D. and Archer, M. C.**, Kinetics of nitrosamine formation in the presence of micelle-forming surfactants, *J. Natl. Cancer Inst.*, 58, 409, 1977.

48. **Collins-Thompson, D. L., Sen, N. P., Aris, B., and Schwinghamer, L.**, Nonenzymic *in vitro* formation of nitrosamines by bacteria isolated from meat products, *Can. J. Microbiol.*, 18, 1968, 1972.

49. **Yang, H. S., Okun, J. D., and Archer, M. C.**, Nonenzymatic microbial acceleration of nitrosamine formation, *J. Agric. Food Chem.*, 25, 1181, 1977.

50. **Suzuki, K. and Mitsuoka, T.**, N-Nitrosamine Formation by Intestinal Bacteria, IARC Sci. Publ., No. 57, International Agency for Research on Cancer, Lyon, 1984, 275.

51. **Mirvish, S. S., Wallcave, L., Eagen, M., and Shubik, P.**, Ascorbate-nitrite reaction: possible means of blocking the formation of carcinogenic N-nitroso compounds, *Science*, 177, 65, 1972.

52. **Newmark, H. L. and Mergens, W. J.**, Blocking nitrosamine formation using ascorbic acid and α-tocopherol, in *Gastrointestinal Cancer: Endogenous Factors*, Banbury Report No. 7, Bruce, W. R., Correa, P., Lipkin, M., Tannenbaum, S. R., and Wilkins, T., Eds., Cold Spring Harbor Laboratory, N. Y., 1981, 285.

53. **Walters, C. L., Edwards, M. V., Elsey, F. S., and Martin, M.**, The effect of antioxidants on the production of volatile nitrosamines during the frying of bacon, *Lebensm. Unters. Forsch.*, 162, 377, 1976.

54. **Fiddler, W., Pensabene, J. W., Piotrowski, E. G., Doerr, R. C., and Wassermann, A. E.**, Use of sodium ascorbate or erythorbate to inhibit formation of N-nitrosodimethylamine in frankfurters, *J. Food Sci.*, 38, 1084, 1973.

55. **Sen, N. P., Donaldson, B., Seaman, S., Iyengar, J. R., and Miles, W. F.**, Inhibition of nitrosamine formation in fried bacon by propyl gallate and L-ascorbyl palmitate, *J. Agric. Food Chem.*, 24, 397, 1976.

56. **Gray, J. I. and Dugan, L. R., Jr.**, Inhibition of N-nitrosamine formation in model food systems, *J. Food Sci.*, 40, 981, 1975.

57. **Mergens, W. J., Kamm, J. J., Newmark, H. L., Fiddler, W., and Pensabene, J.**, Alpha-Tocopherol: Uses in Preventing Nitrosamine Formation, IARC Sci. Publ., No. 19, International Agency for Research on Cancer, Lyon, 1978, 199.

58. **Gray, J. I., Reddy, S. K., Price, J. F., Mandagere, A., and Wilkens, W. F.**, Inhibition of N-nitrosamines in bacon, *Food Technol.*, 36, 39, 1982.

59. **Ohshima, H., Bereziat, J. C., and Bartsch, H.,** Measurement of Endogenous N-Nitrosation in Rats and Humans by Monitoring Urinary and Faecal Excretion of N-Nitrosamino Acids, IARC Sci. Publ., No. 41, International Agency for Research on Cancer, Lyon, 1982, 397.

60. **Pignatelli, B., Bereziat, J. C., O'Neill, I. K., and Bartsch, H.,** Catalytic Role of some Phenolic Substances in Endogenous Formation of N-Nitroso Compounds, IARC Sci. Publ., No. 41, International Agency for Research on Cancer, Lyon, 1982, 413.

61. **Stitch, H. F., Dunn, B. P., Pignatelli, B., Ohshima, H., and Bartsch, H.,** Dietary Phenolics and Betel Nut Extracts as Modifiers of N-Nitrosation in Rat and Man, IARC Sci. Publ., No. 57, International Agency for Research on Cancer, Lyon, 1984, 213.

62. **Kawabata, T., Uibu, J., Ohshima, H., Matsui, M., Hamano, M., and Tokiwa, H.,** Occurrence, Formation and Precursors of N-Nitroso Compounds in the Japanese Diet, IARC Sci. Publ., No. 31, International Agency for Research on Cancer, Lyon, 1980, 481.

63. **Mirvish, S., Nagel, D. L., and Sams, J.,** Methyl- and ethylnitrosocyanamide. Some properties and reactions, *J. Org. Chem.*, 38, 1325, 1973.

64. **Archer, M. C., Clark, S. D., Thilly, J. E., and Tannenbaum, S. R.,** Environmental nitroso compounds: reaction of nitrite with creatine and creatinine, *Science*, 174, 1341, 1971.

65. **Mirvish, S. S., Cairnes, D. A., Hermes, N. H., and Raha, C. R.,** Creatinine: a food component that is nitrosated-denitrosated to yield methylurea, *J. Agric. Food Chem.*, 30, 824, 1982.

66. **Weisburger, J. H., Marquardt, H., Hirota, N., Mori, H., and Williams, G. M.,** *J. Natl. Cancer Inst.*, 64, 163, 1980.

67. **Gray, J. I.,** Formation of N-nitroso compounds in foods, in *N-Nitroso Compounds*, ACS Symp. Series 174, Scanlan, R. A. and Tannenbaum, S. R., Eds., American Chemical Society, Washington, D. C., 1981, 165.

68. **Sakshaug, J., Soegnen, E., Hansen, M. A., and Koppang, N.,** Dimethylnitrosamine; its hepatotoxic effect in sheep and its occurrence in toxic batches of herring meal, *Nature (London)*, 206, 1261, 1965.

69. **Sen, N. P.,** Nitrosamines, in *Safety of Foods*, 2nd ed., Graham, H. D., Ed., AVI Publishing, Westport, Conn., 1980, 319.

70. **Sen, N. P., Seaman, S., and Miles, W. F.,** Volatile nitrosamines in various cured meat products: effect of cooking and recent trends, *J. Agric. Food Chem.*, 27, 1354, 1979.

71. **Sen, N. P., Seaman, S., and McPherson, M.,** Further Studies on the Occurrence of Volatile and Nonvolatile Nitrosamines in Foods, IARC Sci. Publ., No. 31, International Agency for Research on Cancer, Lyon, 1980, 457.

72. **Havery, D. C., Fazio, T., and Howard, J. W.,** Trends in Levels of N-Nitroso pyrrolidine in Fried Bacon, IARC Sci. Publ., No. 19, International Agency for Research on Cancer, Lyon, 1978, 305.

73. **Nitrite Safety Council, U.S.,** A survey of nitrosamines in sausages and drycured meat products, *Food Technol.*, 34, 45, 1980.

74. **Dunn, B. P. and Stitch, H. F.,** Determination of free and protein-bound nitrosoproline in nitrite-cured meat products, *Food Chem. Toxicol.*, 22, 609, 1984.

75. **Eisenbrand, G., Janzowski, C., and Preussmann, R.,** Analysis, formation and occurrence of volatile and nonvolatile N-nitroso compounds: recent results, in *Proc. 2nd Int. Symp. Nitrite Meat Prod.*, Tinbergen, B. J. and Krol, B., Eds., Pudoc, Wageningen, 1977, 155.

76. **Sen, N. P. and Seaman, S.,** On-Line Combination of HPLC-Total N-Nitroso Determination Apparatus for the Determination of N-Nitrosamides and other N-Nitroso Compounds, and some Recent Data on the Levels of N-Nitrosoproline in Foods and Beverages, IARC Sci. Publ., No. 57, International Agency for Research on Cancer, Lyon, 1984, 137.

77. **Eisenbrand, G., Spiegelhalder, B., Janzowski, C., Kann, J., and Preussmann, R.,** Volatile and Nonvolatile N-Nitroso Compounds in Foods and Other Environmental Media, IARC Sci. Publ., No. 19, International Agency for Research on Cancer, Lyon, 1978, 311.

78. **Pensabene, J. W., Feinberg, J. I., Piotrowski, E. G., and Fiddler, W.,** Occurrence and determination of N-nitrosoproline and N-nitrosopyrrolidine in cured meat products, *J. Food Sci.*, 44, 1700, 1979.

79. **Kimoto, W. I., Pensabene, J. W., and Fiddler, W.,** The isolation and identification of N-nitrosothiazolidine in fried bacon, *J. Agric. Food Chem.*, 30, 757, 1982.

80. **Pensabene, J. W. and Fiddler, W.,** N-Nitrosothiazolidine in cured meat products, *J. Food Sci.*, 48, 1870, 1983.

81. **Gough, T. A.,** An Examination of Some Foodstuffs for Trace Amounts of Volatile Nitrosamines Using the Thermal Energy Analyzer, IARC Sci. Publ., No. 19, International Agency for Research on Cancer, Lyon, 1978, 297.

82. **Havery, D. C., Fazio, T., and Howard, J. W.,** Survey of Cured Meat Products for Volatile N-Nitrosamines: Comparison of Two Analytical Methods, IARC Sci. Publ., No. 19, International Agency for Research on Cancer, Lyon, 1978, 41.

83. **Dennis, M. J., Massey, R. C., and McWeeny, D. J.,** The effect of oxygen on nitrosamine formation, *Z. Lebensm. Unters. Forsch.*, 174, 114, 1982.

84. **Walters, C. L., Hart, R. J., and Perse, S.,** The possible role of lipid pseudonitrosites in nitrosamine formation in fried bacon, *Z. Lebensm. Unters. Forsch.*, 168, 177, 1979.
85. **Gray, J. I., Skrypec, D. J., Mandagere, A. K., Booren, A. M., and Pearson, A. M.,** Further Factors Influencing N-Nitrosamine Formation in Bacon, IARC Sci. Publ., No. 57, International Agency for Research on Cancer, Lyon, 1984, 301.
86. **Mottram, D. S., Patterson, R. L. S., Edwards, R. A., and Gough, T. A.,** The preferential formation of volatile N-nitrosamines in the fat of fried bacon, *J. Sci. Food Agric.*, 28, 1025, 1977.
87. **Spiegelhalder, B., Eisenbrand, G., and Preussmann, R.,** Occurrence of Volatile Nitrosamines in Food: A Survey of the West German Market, IARC Sci. Publ., No. 31, International Agency for Research on Cancer, Lyon, 1980, 467.
88. **Sen, N. P., Schwinghamer, L. A., Donaldson, B. A., and Miles, W. F.,** N-Nitrosodimethylamine in fish meal, *J. Agric. Food Chem.*, 20, 1280, 1972.
89. **Preussmann, R.,** Occurrence and Exposure to N-Nitroso Compounds and Precursors, IARC Sci. Publ., No. 57, International Agency for Research on Cancer, Lyon, 1984, 3.
90. **Fazio, T. and Havery, D. C.,** Volatile N-Nitrosamines in Direct Flame Dried Processed Foods, IARC Sci. Publ., No. 41, International Agency for Research on Cancer, Lyon, 1982, 277.
91. **Havery, D. C., Hotchkiss, J. H., and Fazio, T.,** Nitrosamines in malt and malt beverages, *J. Food Sci.*, 46, 501, 1981.
92. **Scanlan, R. A., Barbour, J. F., Hotchkiss, J. H., and Libbey, L. M.,** N-Nitrosodimethylamine in beer, *Food Cosmet. Toxicol.*, 18, 27, 1980.
93. **Sen, N. P., Seaman, S., and Tessier, L.,** Comparison of two analytical methods for the determination of dimethylnitrosamine in beer and ale, and some recent results, *J. Food Safety*, 4, 243, 1982.
94. **Weston, R. J.,** Nitrosamine content of New-Zealand beer and malted barley, *J. Sci. Food Agric.*, 34, 1005, 1093.
95. **Goff, E. U. and Fine, D. H.,** Analysis of volatile N-nitrosamines in alcoholic beverages, *Food Cosmet. Toxicol.*, 17, 569, 1979.
96. **Havery, D. C., Hotchkiss, J. H., and Fazio, T.,** A rapid method for the determination of volatile nitrosamines in nonfat dry milk, *J. Dairy Sci.*, 65, 182, 1981.
97. **Lakritz, L. and Pensabene, J. W.,** Survey of fluid and nonfat dry milks for N-nitrosamines, *J. Dairy Sci.*, 64, 371, 1981.
98. **Wainright, T., Slack, P. T., and Long, D. E.,** N-Nitrosodimethylamine Precursors in Malt, IARC Sci. Publ., No. 41, International Agency for Research on Cancer, Lyon, 1982, 71.
99. **Mangino, M. M. and Scanlan, R. A.,** Rapid Formation of NDMA from the Alkaloid Gramine: a Naturally-Occurring Tertiary Amine Precursor of NDMA in Malt, IARC Sci. Publ., No. 57, International Agency for Research on Cancer, Lyon, 1984, 337.
100. **Gray, J. I., Irvine, D. M., and Kakuda, Y.,** Nitrate and N-nitrosamines in cheese, *J. Food Prot.*, 42, 263, 1979.
101. **Groenen, P. J., Luten, J. B., Dhont, J. H., De Cock-Bethbeder, M. W., Prins, L. A., and Vreekin, J. W.,** IARC Sci. Publ., No. 41, International Agency for Research on Cancer, Lyon, 1982, 99.
102. **Lu, S.-X., Li, M.-X., Ji, M.-Y., Wang, Y.-L., and Huang, L.,** A new N-nitroso compound, N-3-methylbutyl-N-1-methylacetonylnitrosamine, in corn-bread inoculated with fungi, *Sci. Sin.*, 22, 601, 1979.
103. **Hoffmann, D., Brunnemann, K. D., Adams, J. D., Rivenson, A., and Hecht, S. S.,** N-Nitrosamines in tobacco carcinogenesis, in *Nitrosamines and Human Cancer*, Banbury Report 12, Magee, P. N., Ed., Cold Spring Harbor Laboratory, N. Y., 1982, 211.
104. **Maki, T., Tamura, Y., Shimamura, Y., and Navi, Y.,** Estimate of volatile nitrosamines in Japanese food, *Bull. Environ. Contam. Toxicol.*, 25, 257, 1980.
105. **Stephany, R. W. and Schuller, P. L.,** Daily dietary intakes of nitrate, nitrite, and volatile N-nitrosamines in the Netherlands using the duplicate portion sampling technique, *Oncology*, 37, 203, 1980.
106. **Gough, T. A., Webb, K. S., and Coleman, R. F.,** Estimate of the volatile nitrosamine content of UK food, *Nature (London)*, 272, 161, 1978.
107. **Kawabata, T., Matsui, M., Ishibashi, T., and Hamano, M.,** Analysis and Occurrence of Total N-Nitroso Compounds in Japanese Diet, IARC Sci. Publ., No. 57, International Agency for Research on Cancer, Lyon, 1984, 25.
108. **Wagner, D. A., Shuker, D. E. G., Bilmazes, C., Obiedzinski, M., Young, V. R., and Tannenbaum, S. R.,** Modulation of Endogenous Synthesis of N-Nitrosoproline by Dietary Factors, IARC Sci. Publ., No. 57, International Agency for Research on Cancer, Lyon, 1984, 223.
109. **Tsuda, M., Kakijoe, T., Hirayama, T., and Sugimura, T.,** Study on Analysis of N-Nitrosamines in Human Urine by GC-TEA, IARC Sci. Publ., No. 57, International Agency for Research on Cancer, Lyon, 1984, 87.

110. **Reed, P. I., Smith, P. L. R., Summers, K., Walters, C. L., Bartholomew, B. A., Hill, M. J., Vennit, S., House, F. R., Horning, D., and Bonjour, J. P.,** Ascorbic Acid Treatment in Achlorhydric Subjects — Effect on Gastric Juice Nitrite and N-Nitroso Compound Formation, IARC Sci. Publ., No. 57, International Agency for Research on Cancer, Lyon, 1984, 977.

111. **Bartsch, H., Ohshima, H., Muñoz, N., Crespi, M., Cassale, V., Ramazotti, V., Lehton, A.,Inberg, M., Aitio, A., Lambert, R., Minaire, Y., Forichon, J., Tulinius, H., and Walters, C. L.,** In-vivo Nitrosation, Precancerous Lesions and Cancers of the Gastrointestinal Tract. On-going Studies and Preliminary Results, IARC Sci. Publ., No. 57, International Agency for Research on Cancer, Lyon, 1984, 957.

112. **Fine, D., Ross, R., Rounbehler, D. P., Silvergleid, A., and Song, L.,** Formation *in vivo* of volatile N-nitrosamines in man after ingestion of cooked bacon and spinach, *Nature (London)*, 265, 753, 1977.

113. **Kowalski, B., Miller, C. T., and Sen, N. P.,** Studies on the in vivo Formation of Nitrosamines in Rats and Humans after Ingestion of Various Meals, IARC Sci. Publ., No. 31, International Agency for Research on Cancer, Lyon, 1980, 609.

114. **Melikian, A. A., LaVoie, E. J., Hoffmann, D., and Wynder, E. L.,** Volatile nitrosamines: analysis in breast fluid and blood of non-lactating women, *Food Cosmet. Toxicol.*, 19, 757, 1981.

115. **Yamamoto, M., Yamada, T., and Tanimura, A.,** Volatile nitrosamines in human blood before and after ingestion of a meal containing high concentrations of nitrate and secondary amines, *Food Cosmet. Toxicol.*, 18, 297, 1980.

116. **Garland, W., Holowaschenko, H., Kuenzig, W., Norkus, E. P., and Conney, A. H.,** A high resolution mass spectrometry assay for N-nitrosodimethylamine in human plasma, in *Nitrosamines in Human Cancer*, Banbury Report 12, Magee, P. N., Ed., Cold Spring Harbor Laboratory, N. Y., 1982, 183.

117. **Swann, P.,** The Effect of Ethanol on Nitrosamine Metabolism and Distribution, IARC Sci. Publ., No. 57, International Agency for Research on Cancer, Lyon, 1984.

118. **Lakritz, L., Simenhoff, M. L., Dunn, S. R., and Fiddler, W.,** N-Nitrosodimethylamine in human blood, *Food Cosmet. Toxicol.*, 18, 77, 1980.

119. **Yamamoto, M., Iwata, R., Ishiwata, H., Yamada, T., and Tanimura, T.,** Determination of volatile nitrosamine levels in foods and estimation of their daily intake in Japan, *Food Chem. Toxicol.*, 22, 61, 1984.

120. **Sen, N. P.,** unpublished data.

121. **Sen, N. P., Seaman, S. W., and Baddoo, P. A.,** N-Nitrosothiazolidine and nonvolatile N-nitroso compounds in foods, *Food Technol.*, 39, 84, 1985.

122. **Sen, N. P., Baddoo, P. A., Seaman, S. W., and Iyengar, J. R.,** N-Nitrosothiazolodine-4-carboxylic acid and N-nitrosothiazolodine in smoked meats and fish, presented at the 45th Annual Meeting of the Food Technologists, Atlanta, GA, June 9-12, 1985, Abstr. No. 304

123. **Mandagere, A. K., Ilkins, W. G., Gray, J. I., Booren, A. M., and Pearson, A. M.,** Formation of N-nitrosothiazolidine and its carboxylic acid in smoked cured meat systems, presented at the 45th Annual Meeting of the Food Technologists, Atlanta, GA, June 9-12, 1985, Abstr. No. 305.

124. **Pensabene, J. W. and Fiddler, W.,** Formation and inhibition of N-nitrosothiazolidine in bacon, *Food Technol.*, 39, 91, 1985.

125. **Hotchkiss, J. H. and Vecchio, A. J.,** Nitrosamines in fried-out bacon fat and its use as a cooking oil, *Food Technol.*, 39, 67, 1985.

126. **Hotchkiss, J. H., Vecchio, A. J., and Ross, H. D.,** N-Nitrosamine formation in fried-out bacon fat: Evidence for nitrosation by lipid-bound nitrite, *J. Agric. Food Chem.*, 33, 5, 1985.

Chapter 10

DIET, NUTRITION, AND CANCER: DIRECTIONS FOR RESEARCH*

Sushma Palmer** and Kulbir Bakshi

TABLE OF CONTENTS

*   This chapter summarizes the final report of the National Research Council Committee on Diet, Nutrition, and
    Cancer. The committee was chaired by Clifford Grobstein and John Cairns was Vice Chairman. Members were
    Selwyn A. Broitman, T. Colin Campbell, Joan D.Gussow, Laurence N. Kolonel, David Kritchevsky, Walter
    Mertz, Anthony B. Miller, Michael J. Prival, Thomas Slaga, and Lee Wattenberg. Takashi Sugimura was an
    advisor to the committee. The study was conducted pursuant to Contract No. NO1-CP-05603 with the National
    Cancer Institute, Bethesda, Md.
** The authors and the Committee on Diet, Nutrition, and Cancer wish to acknowledge the assistance of numerous
    scientists who contributed to the preparation of the report.

I. INTRODUCTION

Over the past half century experiments in laboratory animals, and more recently studies in human populations have generated an abundance of data pertaining to the role of dietary components in the etiology and prevention of cancer. The Committee on Diet, Nutrition, and Cancer (the committee) of the National Research Council recently analyzed this literature and published a report.[10,11] The major conclusions and interim dietary guidelines proposed by the committee are summarized in Tables 1 to 4.

As summarized in Tables 1 to 3, the analysis of the literature pointed to several dietary patterns that appear to enhance carcinogenesis (e.g., a high-fat diet) and others that seem to have a protective effect (e.g., frequent consumption of carotene-rich and *Brassica* family vegetables). On the basis of this knowledge, the committee proposed certain interim dietary guidelines to lower cancer risk (see Table 4). It also highlighted the limitations of knowledge concerning the precise role of individual foods and food constituents in carcinogenesis, particularly their mechanisms of action. This chapter is concerned solely with the directions and strategies for future research to enhance our understanding of the effect of diet on carcinogenesis, the subject of the second report of the Committee on Diet, Nutrition, and Cancer.[12]

Defining profitable directions for research poses a formidable challenge because it is impossible to predict exactly where major discoveries will be made, and thus any attempt to stipulate a particular sequence for research might stifle creativity. Nevertheless, delineating a flexible conceptual framework for research may be useful in accomplishing a defined practical objective such as lowering cancer risk. This underlying philosophy guided the committee in its preparation of directions for research. As explained below, further clues were obtained through an understanding of commonalities in the sequence of discoveries in some branches of nutrition research as well as the study of carcinogenesis.

Many diseases now known to be associated with particular dietary imbalances (e.g., vitamin deficiencies) or specific toxic contaminants in foods (e.g., mycotoxins) were initially only crudely linked to some dietary peculiarity (e.g., a lack of fresh foods or the consumption of moldy grains) (for example, see King,[6] Kraybill and Shimkin,[7] Keen and Martin[5]). Crude additions to or eliminations from the diet were successful in alleviating many such dietary imbalances or mycotoxicosis, long before animal models that simulated those diseases were developed or the biochemical pathways to elucidate their mechanisms and explain their etiology were identified (for example, see White et al.,[16] Wogan[17]). Similarly, in cancer research, epidemiological associations of cancer with exposure to certain chemical mixtures such as chimney soot and the prevention of the cancer by eliminating such exposure, far preceded the identification of the active carcinogens or the discovery of their mechanism of action (for example, see Pott[13]). These historical scenarios may not be indicative of the precise course of research needed to unravel interactions between a complex mixture like diet and a multifactorial disease like cancer, but they may provide clues as to what to expect.

It was apparent to the Committee on Diet, Nutrition, and Cancer that by relying solely on the published literature only those gaps in knowledge might be identified that could be deduced from the already published literature. Therefore, to avoid overlooking new avenues of research that might have an important bearing on the relationship between diet and cancer and to obtain insights from other investigators in the field, input was sought from an advisory panel of nearly 100 distinguished scientists knowledgeable about cancer, nutrition, and related areas. The effort to draw conclusions about areas that deserved particular emphasis resulted in focus on five major categories of research: diet and the mechanisms of carcinogenesis, epidemiological methodology, laboratory methodology, dietary macroconstituents, and dietary microconstituents. Seven broad sets of strategic objectives and general recommendations that emerged from a synthesis and refinement of ideas generated by the panelists and the committee are summarized below.

Table 1
DIET, NUTRITION, AND CANCER: GENERAL CONCLUSIONS*

1. The differences in the rates at which various cancers occur in different human populations are often correlated with differences in diet. The likelihood that some of these correlations reflect causality is strengthened by laboratory evidence that similar dietary patterns and components of food also affect the incidence of certain cancers in animals.

2. In general, the evidence suggests that some types of diets and some dietary components (e.g., high-fat diets or the frequent consumption of salt-cured, salt-pickled, and smoked foods) tend to increase the risk of cancer, whereas others (e.g., low-fat diets or the frequent consumption of certain fruits and vegetables) tend to decrease it. The mechanisms responsible for these effects are not fully understood.

3. The evidence suggests that cancers of most major sites are influenced by dietary patterns. However, the data are not sufficient to quantitate the contribution of diet to the overall cancer risk or to determine the percent reduction that might be achieved by dietary modifications.

Table 2
DIET, NUTRITION, AND CANCER: CONCLUSIONS ON DIETARY MACROCONSTITUENTS*

Total Caloric Intake — Neither the epidemiological studies nor the experiments in animals permit a clear interpretation of the specific effect of total caloric intake on the risk of cancer. Nonetheless, the studies conducted in animals show that a reduction in the total food intake decreases the age-specific incidence of cancer. The evidence is less clear for human beings.

Fats and Lipids — Of all the dietary components studied, the combined epidemiological and experimental evidence is most suggestive for a causal relationship between high fat intake and increased occurrence of cancer. The relationship between dietary cholesterol and carcinogenesis is not clearly understood.

Protein — The evidence from both epidemiological and laboratory studies suggests that high protein intake *may* be associated with an increased risk of cancers at certain sites. However, because of the relative paucity of data on protein compared to fat, and the strong correlation between the intakes of fat and protein in the U.S. diet, no firm conclusions can be drawn about an independent effect of protein.

Carbohydrates — Information concerning the role of carbohydrates in the development of cancer is extremely limited and inconclusive.

Dietary Fiber — There is no conclusive evidence to indicate that *total dietary fiber* exerts a protective effect against colorectal cancer in humans. Both epidemiological and laboratory reports suggest that if there is such an effect, specific components of fiber, rather than total fiber, are more likely to be responsible.

Alcoholic Beverages — In some countries, including the U.S., excessive beer drinking has been associated with an increased risk of colorectal cancer, especially rectal cancer. There is evidence that excessive alcohol consumption causes hepatic injury and cirrhosis, which in turn may lead to the formation of hepatomas. Excessive consumption of alcoholic beverages and cigarette smoking appear to act synergistically to increase the risk for cancers of the mouth, larynx, esophagus, and the respiratory tract.

II. TYPES OF RECOMMENDATIONS

The report contains both general and specific recommendations; however, the complex interrelationships among them precluded any attempt to set priorities.

Three kinds of general recommendations were derived, the first set from gaps in the current state of knowledge about the effects of specific dietary components. Where conclusions were uncertain, simply because information was incomplete, recommendations were directed toward supplementing the data base. Where the uncertainty stemmed from the imprecision of the data, there were recommendations for methodological improvement. Where several causal factors seemed to be operating, there were recommendations that their interactions be studied. Such recommendations only highlighted what would be obvious to an experienced investigator.

A second kind of recommendation arose from the fact that the etiology of carcinogenesis is complex, involving several combinations of dietary and other components, complex steps in cancer progression, and a multiplicity of cancer types. This complexity led to suggestions

* From Diet, Nutrition, and Cancer, National Research Council, National Academy Press, Washington, D. C., 1982.

Table 3
DIET, NUTRITION, AND CANCER: CONCLUSIONS ON MICROCONSTITUENTS*

Vitamins — The laboratory evidence suggests that *vitamin A* itself and many of the *retinoids* are able to suppress chemically induced tumors. The epidemiological evidence is sufficient to suggest that *foods rich in carotenes* or *vitamin A* are associated with a reduced risk of cancer. Limited evidence suggests that *vitamin C* can inhibit the formation of some carcinogens and that the consumption of *vitamin C-containing foods* is associated with a lower risk of cancers of the stomach and esophagus. The data are not sufficient to permit any firm conclusions to be drawn about the effect of *vitamin E* or the *B vitamins* on cancer in humans.

Fruits and Vegetables — There is sufficient epidemiological evidence to suggest that consumption of certain vegetables, especially *carotene-rich vegetables* (i.e., dark green and deep yellow) and *vegetables* of the *Brassica* genus (e.g., cabbage, broccoli, cauliflower, and brussels sprouts) is associated with a lower incidence of cancer at several sites in humans. A number of nonnutritive and nutritive compounds that are present in these vegetables also inhibit carcinogenesis in laboratory animals. Investigators have not yet established which, if any, of these compounds may is responsible for the protective effect observed in epidemiological studies.

Minerals — Both the epidemiological and laboratory studies suggest that *selenium* may offer some protection against the risk of cancer. However, firm conclusions cannot be drawn from the limited evidence. The data concerning dietary exposure to *iron, zinc, copper, molybdenum, iodine, arsenic, cadmium,* and *lead* are insufficient and provide no basis for conclusions about the association of these elements with carcinogenesis.

Food Additives, Contaminants, and Naturally-Occurring Carcinogens — The increasing use of *food additives* does not appear to have contributed significantly to the overall risk of cancer for humans. However, this lack of detectable effect may be due to their lack of carcinogenicity, to the relatively recent use of many of these substances, or to the inability of epidemiological techniques to detect the effects of additives against the background of common cancers from other causes. A number of *environmental contaminants* (e.g., some organochlorine pesticides, polychlorinated biphenyls, and polycyclic aromatic hydrocarbons) cause cancer in laboratory animals. The committee found no epidemiological evidence to suggest that these compounds individually make a major contribution to the risk of cancer in humans. However, the possibility that they may act synergistically and may thereby create a greater carcinogenic risk cannot be excluded. Certain *naturally-occurring contaminants* in food (e.g., aflatoxin and N-nitroso compounds), and *nonnutritive constituents* (e.g., hydrazines in mushrooms) are carcinogenic in animals and pose a potential risk of cancer to humans. These and other compounds thus far shown to be carcinogenic in animals have been reported to occur in the average U.S. diet in small amounts; however, there is no evidence that any of these substances *individually* makes a major contribution to the total risk of cancer in the U.S. This lack of sufficient data should not be interpreted as an indication that these or other compounds subsequently found to be carcinogenic do not present a hazard.

Mutagens in Food — Most *mutagens* detected in foods have not been adequately tested for carcinogenic activity. Although substances that are mutagenic are *suspect carcinogens,* it is not yet possible to assess whether such mutagens are likely to contribute significantly to the incidence of cancer in the U.S.

Table 4
DIET, NUTRITION, AND CANCER: INTERIM DIETARY GUIDELINES**

- Reduce intake of both saturated and unsaturated fats, from ~40% to approximately 30% of total calories.
- Include fruits, vegetables, and whole-grain cereal products in daily diet, especially citrus fruits and carotene-rich and cabbage family vegetables. Avoid high-dose supplements of individual nutrients.
- Minimize consumption of cured, pickled, and smoked foods.
- Drink alcohol only in moderation.

for long-term, multifaceted studies that are necessarily large and therefore unavoidably expensive. Such studies are likely to be most informative when they can be conducted under especially favorable circumstances. For example, high-risk human populations may be studied under circumstances that permit simultaneous study of multiple risk factors. Information from such studies could be supplemented with laboratory data derived from a suitable animal model.

A third kind of recommendation calls attention to the need for behavioral and social

---

* From Diet, Nutrition, and Cancer, National Research Council, National Academy Press, Washington, D.C., 1982.
** Proposed by the Committee on Diet, Nutrition, and Cancer of the National Academy of Sciences as sound nutritional practices that are likely to reduce the risk of cancer (National Research Council, 1982).

studies. This stemmed from the assessment of the literature that indicated that most common cancers appear to be influenced by diet, suggesting that, to a certain extent, individuals may be able to influence their chances of getting cancer. However, knowledge that a certain exposure strongly influences cancer is apparently not sufficient to convince people to modify their behavior. For example, it is clear that simply demonstrating a causal connection between smoking and lung cancer has not eliminated the smoking habit. Therefore, the need for social and behavioral research to supplement research in the area of physiology, pathology, cytology, nutrition, and biochemistry was emphasized.

The report deals primarily with research on diet, nutrition, and carcinogenesis. Even though it would seem that research on diet and carcinogenesis can progress without awaiting further discoveries in basic mechanisms, the committee was aware of the need to press forward simultaneously with fundamental investigations, particularly taking advantage of new opportunities afforded by recent advances in cellular and molecular biology (e.g., Logan and Cairns[8]). Such fundamental research is likely to improve not only an understanding of the impact of diet and nutrition on carcinogenesis but also the ability to address many aspects of the prevention and treatment of cancer.

III. STRATEGIC OBJECTIVES AND CORRESPONDING RECOMMENDATIONS

The committee operated on the premise that research on diet and carcinogenesis should be viewed in a logical but flexible conceptual framework that should encompass numerous sources of data, i.e., surveys to monitor exposure, epidemiological studies, carcinogenesis bioassays in animals, short-term tests for genotoxicity, short-term in vivo bioassays to detect early biological indications of carcinogenesis, and studies designed to elucidate metabolic pathways or pathogenic mechanisms. The following general recommendations for epidemiological and laboratory research and the seven strategic objectives that follow illustrate this principle.

A. General Recommendations for Epidemiological Research

There is probably considerable interaction among the many components of the diet. Therefore, some potentially harmful substances may be "neutralized" by other dietary ingredients. For this reason, the committee ascertained that there is a need to evaluate the interrelationships among calories, protein, and fat (and its various components) and their effect on, for example, breast and colorectal cancer.

Simultaneously, the opportunity should be taken to evaluate (1) the effects of hormonal status on breast cancer and (2) the effects of fiber, its components, and various micronutrients, especially vitamins and possible inhibitors in vegetables of the genus *Brassica*, on the etiology of colorectal cancer.

Research on interrelationships among macroconstituents should be designed so that it is possible to determine the overall effect of different groups of foods and not just individual foods or nutrients.

The completed dietary studies on breast and colon cancer should be extended to examine other possibly diet-associated cancers that have been correlated with breast and colon cancer. In addition to the on-going investigations of prostate cancer, studies should be conducted on endometrial, ovarian, pancreatic, and renal cancers.

More frequent monitoring of food intake, especially changes in intake of macronutrients in the average diet, is essential.

Reliable data bases for food composition should be developed for the analysis of macronutrients. It is especially important that such data bases contain more information on the fiber content of each food and on the chemical composition of each type of fiber.

It is now essential to give high priority to some long-term cohort studies that will test hypotheses about macroconstituents and cancer.

Carefully planned intervention studies, involving changes in the macronturient content of the diet, should be conducted in humans. Such studies may be the only way to gain an understanding of their relative effects and the interrelationships among macroconstituents. In such studies, it may be conceptually easier to plan for the addition of constituents, e.g., specific types of fiber, to the diet. However, we should not overlook the need to evaluate the effect of reducing dietary fat by consumption of foods low in fat.

B. General Recommendations for Laboratory Research

Laboratory techniques for investigations of diet and carcinogenesis require innovation. Techniques should be developed to identify early biological or biochemical indicators of carcinogenesis. This recommendation is based on the concern that where neoplasia is used as the sole end point, investigators are severely limited by the long latency period between exposure and expression.

For animal experiments, the first priority is standardization of methodology, as well as development of suitable animal models. For example, the epidemiological data linking specific dietary components to cancers of the prostate, pancreas, and endometrium are limited. An expansion of the experimental data base is required in order to put these data into proper perspective.

In general, better laboratory methods based on mechanism of action need to be devised to detect carcinogens, promoters, cocarcinogens, and inhibitors in food and to extrapolate estimates of risk to human health derived from laboratory studies.

C. Stragetic Objectives

1. (Further) Identification of the Foods and of the Dietary Macro- and Microconstituents that Alter the Risk of Cancer, and Elucidation of their Mechanisms of Action

This objective derived directly from the assessment of the literature that had resulted in identification of four categories of dietary constituents that are likely to affect the risk of cancer. These categories were saturated and unsaturated fat; certain fruits, vegetables, and whole grain cereals; smoked, cured, and pickled foods; and alcoholic beverages.[10] The committee recommended that when the epidemiological and experimental evidence associating particular dietary components with cancer risk is sufficiently convincing, studies should be undertaken to identify the specific active dietary constituents and their mechanisms of action. Information from such studies would be useful in refining the interim dietary guidelines in the first report.[10] The following illustrates the types of epidemiological and laboratory research recommended by the committee to accomplish this objective.

a. Dietary Macroconstituents

The mechanism(s) by which a high-fat diet increases the incidence of certain cancers should be elucidated. In this context, the point at which dietary fat exerts its precise effects during the prepromotional, promotional, tumor development, and metastatic stages needs to be systematically investigated. Furthermore, the relative roles of the level and type of fat (e.g., essential fatty acids) in all phases of tumor formation should be studied.

The mechanism underlying the reported association between low blood cholesterol and neoplasia needs to be determined. For example, it would be helpful to know whether hypocholesterolemic individuals are at high risk only if they consume a high-fat diet and whether lowering fat intake in such individuals will reduce their risk. Further analysis of existing data on humans may help to answer these questions.

Other macroconstituents also need attention: for example, studies are needed to (1) investigate the effect of protein on different stages of carcinogenesis and the mechanism underlying this effect and (2) to further study the relationship between total caloric intake and cancer in humans, and in the laboratory, to determine the mechanism for the putative effect of caloric intake on carcinogenesis.

It would be helpful to investigate the influence of fiber on cancers other than colorectal cancer and the structure-function relationships of fiber (e.g., its pentose content and its bile-acid binding capacity) in tumor formation.

The influence of different alcoholic beverages (e.g., wine, beer, whisky, or liqueurs) on esophageal, gastric, and other cancers needs to be investigated. Furthermore, studies should be conducted to determine the influence of nonalcoholic components (contaminants) of alcoholic beverages on experimentally induced carcinogenesis, as well as the association between carcinogenesis and nutrient deficiences imposed by excessive alcohol intake.

Experiments should be conducted in animals to evaluate the effects of dietary macro- and microconstituents on the later stages of carcinogenesis.

b. Dietary Microconstituents

Having concluded that frequent consumption of certain fruits and vegetables is associated with a lower cancer risk, it is desirable to identify the constituents of fruits and vegetables that are responsible for the observed reduction in risk associated with their frequent consumption and to define the mechanisms of action of those constituents. In addition, research needs to be conducted to determine the laboratory conditions under which the occurrence of neoplasia can be prevented by microconstituents, especially, vitamins A, C, and E and the carotenes, selenium and possibly other trace minerals, and nonnutritive inhibitors of carcinogenesis that occur naturally in the *Brassica*-family vegetables.

Other nonnutritive microconstituents in the diet also need attention: there is a need to continue to evaluate the carcinogenic potential of suspect compounds in common foods. These compounds include certain mycotoxins, polycyclic aromatic hydrocarbons, and naturally occurring constituents such as flavonoids and methylglyoxal.

c. Food Additives, Contaminants, Mutagens

Furthermore, the search for possible tumor-promoting activity of food additives and contaminants should be pursued. For example, studies should be conducted to examine the tumor-promoting effects of butylated hydroxytoluene (BHT) and the tumor-inhibiting effects of both BHT and butylated hydroxyanisole (BHA) to determine their relevance to humans. These are such widely used additives with known effects in experimental systems that intensive investigation is warranted. An effort should be made to determine the feasibility of epidemiological studies on these widely distributed substances.

Having determined that cooked foods contain many different mutagens, it would be helpful to identify compounds responsible for most of the mutagenic activity in normally prepared foods and beverages.

It is desirable to determine the effects of diet on the endogenous formation of mutagens, such as nitrosamines and fecal and urinary mutagens, and assess the carcinogenicity of such mutagens. Chemical identification of nitrosatable precursors and endogenously produced mutagens should be pursued.

An attempt should be made to investigate the possibility that comutagens and inhibitors of mutagenesis may work through mechanisms that are relevant in vivo. For example, food chemicals that are mutagenic in vitro should be assessed for stability in the GI tract. In some cases, efforts to identify DNA adducts formed in vivo may be useful. At present, such effects observed in in vitro mutagenicity assays may simply be artifacts related to conditions in the assay systems being used.

2. Improvement of the Data Base and the Methodology for Assessing Exposure of Humans to Foods and Dietary Constituents that May Alter the Risk of Cancer

Epidemiological shortcomings presented the greatest obstacles to obtaining definitive data from which precise conclusions could be drawn. Therefore, high priority needs to be given to developing better methods to monitor and quantify dietary exposures in human populations. The resulting improvement in the data base should enable investigators to establish more clearly the relationship of dietary constituents and dietary patterns to the occurrence of cancer.

Improvement is needed both in the techniques for assessing dietary exposure, as well as in the analysis of the data as summarized below.

a. Assessment of Dietary Exposure

New approaches are needed to improve the quality of dietary intake data, especially in relation to long-term dietary patterns, and more effort should be spent in evaluating the validity of dietary methods.

Food composition data bases should be improved with respect to information on both nutritive and nonnutritive constituents and on regional and seasonal variations in composition. Better data are also needed on long-term trends in the composition of average national diets. On a regular basis, cross-sectional information should be collected from representative samples of the U.S. population through surveys such as the Health and Nutrition Examination Surveys (HANES) conducted by the National Center for Health Statistics.[3] The accumulated data can then be correlated with trends in cancer incidence.

The content of fiber components in foods should be determined in order to assess dietary intake more accurately. In addition, the physiological properties of different components of fiber and their effects on the absorption and availability of nutrients should be systematically evaluated in metabolic studies.

Investigations should be expanded to include a determination of the amounts of micro-constituents in various foods and followed by analytical epidemiological studies (case-control or cohort studies) to determine the effects exerted by various levels of microconstituent intake on the occurrence of cancers in humans.

Better measurement of the levels of food additives consumed and the distribution of their intake among different population subgroups should be obtained. In addition, existing food intake data should be used, if possible, to determine the relationship between the levels of food additives produced and the amounts consumed. Such studies are needed in order to assess levels of exposure to both direct and indirect additives. Once populations with different levels of exposure to food additives are identified, epidemiological studies should be conducted to evaluate the effect of these additives on cancer risk.

Efforts should be made to obtain better measurements of the level of consumption levels and the distribution of intake among different population subgroups for carcinogens and mutagens in foods, such as hydrazines in mushrooms, aflatoxins, other mycotoxins, mutagenic flavonoids, and mutagens resulting from cooking. This effort would have to include a study of the patterns and frequencies of household and commercial cooking practices, including the cooking temperature and the duration of cooking for various types of food in which mutagens or carcinogens are produced during heating.

b. Analysis of Data

The analysis of data from epidemiological studies should include an examination of specific foods and food classes, even when the hypotheses pertain to nutrients. Since many of the dietary and nondietary factors in studies on cancer are highly intercorrelated, efforts should be made to explore statistical methods that are less sensitive to collinearity than is multiple logistic regression analysis. Finally, statistical techniques need to be developed to describe more accurately the various forms of interaction among dietary variables.

3. Identification of Markers of Exposure and Early Indicators of the Risk of Cancer

Research should be conducted to identify early biological markers of exposure to chemicals that cause cancer in humans, especially markers that can forecast the emergence of clinical cancer. Studies should also be undertaken to determine if there are biochemical markers that are indicative of long-term nutritional status of humans with regard to dietary constituents.

Short-term test systems should be developed to detect the early effects of dietary initiators to which humans are exposed. Thus, techniques should be developed to identify the presence of carcinogen-DNA adducts and to detect alterations in DNA. In addition, attempts should be made to refine cytogenetic procedures. Attempts should also be made to detect early the impact of compounds suspected of acting as promoters or cocarcinogens in humans.

Techniques should be developed for detecting dietary compounds that have the capacity to inhibit carcinogenesis. These should be applied systematically to identify such inhibitors, and to evaluate their protective effects.

Techniques need to be developed for studying the in vivo formation of carcinogens and mutagens in humans, and for assessing the mutagenic effects of chemicals on human cells in vivo. As such techniques become available, they should be applied to test populations known to be consuming diets that are believed to present a high or a low risk for cancer.

4. Determination and Quantification of the Adverse or Beneficial Effects of the Foods and of the Dietary Macro- and Microconstituents that Affect the Risk of Cancer

Although several dietary constituents are now known to affect carcinogenesis, relatively little is known about the magnitude of this effect. Therefore, effort needs to be made to "quantify" the effects of potentially carcinogenic or inhibitory substances. As outlined below, these studies should include a focus on substances that can damage macromolecules, especially DNA; on those that enhance experimentally induced carcinogenesis, i.e., promoters and cocarcinogens; and on those that can inhibit experimentally induced carcinogenesis.

Methods are needed to study dose-response relationship between dietary constituents in general and nutrients in particular, and tumorigenesis. For example, more discriminating data are needed on the effect of the level and type of fat intake. In the laboratory, it would be important to determine if there is a threshold at which dietary fat begins to exert an effect on carcinogenesis, and in humans to determine the level of fat intake that is associated with the maximum reduction in cancer incidence. Is it 30% of calories, 25%, 20%, or less?

Methods should also be devised and used systematically to evaluate the association between carcinogenesis and the interaction of nutrients with each other and with nonnutritive dietary constituents. For example, the independent effects of the amount and type of protein and of their interaction with other macroconstituents on the incidence of tumors in humans needs to be investigated.

Reliable methods are needed to quantitate alcohol intake. This should be followed by investigation of the effects of different consumption levels on cancer risk.

5. Determination of the Ranges of Optimal Intake of Dietary Macro- and Microconstituents

The committee emphasized that attention should be given to determining ranges of dietary macro- and microconstituents that are optimal not merely for the prevention of deficiency diseases but also for the promotion of other aspects of health, including the reduction of the risk of cancer. For example, it would be useful to establish a dose-response curve for selenium and to define the optimal range of selenium intake, giving special attention to levels that might be needed to achieve a reduction in the risk of certain cancers.

Similarly, more discriminating data are needed on the effects of different amounts and types of fat. Studies should be conducted to answer the following questions: Is the finding that polyunsaturated fat increases tumor incidence in laboratory animals relevant to humans? Is this finding also relevant to the results emerging from intervention trials for cardiovascular disease? What should be the optimal proportion of polyunsaturated, monounsaturated, and saturated fats in the diet?

Efforts should be made to establish dose-response curves for microconstituents (e.g., vitamin A, carotenes, and selenium) that inhibit carcinogenesis.

Consideration needs to be given to studying the incidence of specific cancers in populations consuming large doses of vitamin supplements.

6. Intervention to Reduce the Risk of Cancer

Intervention studies (trials) have many advantages, but for ethical reasons, they will have to be limited to examinations of putative protective factors or of the effects from reduced exposure to risk factors. The committee recommended that such trials be conducted with foods or food constituents believed to be associated with a lower cancer risk, but as outlined below only when a substantial body of data indicates a high likelihood of benefit without discernible risk.

Thus, there is a need for carefully planned intervention studies involving changes in the macronutrient content of the diet. Such studies may be the major avenue for gaining an understanding of the relative effects and the interrelationships among macronutrients. In such studies, it may be conceptually easier to plan for the addition of constituents, e.g., specific types of fiber, to the diet. However, the need to evaluate the effect of reducing dietary fat by consumption of foods low in fat should also be pursued.

When justified by sufficiently definitive data from experimental and/or epidemiological investigations, intervention studies with microconstituents or with foods rich in these substances should be considered. Epidemiological studies, including intervention trials when appropriate, should be conducted to determine if consumption of foods containing high concentrations of these compounds results in a lower incidence of cancer. Several trials of specific microconstituents (e.g., beta carotene) have recently been initiated. However, there is also a great need to study the effects of specific foods and food groups (e.g., dark green and deep yellow vegetables, and those of the genus *Brassica*, i.e., members of the cabbage family), since the biological effectiveness of a food component is probably affected by the presence of other constituents in the diet, and since the effects observed in epidemiological and experimental studies may be due to a mixture of different inhibitors of carcinogenesis. To obtain more definitive data, the randomization procedure in these studies should be based on individuals rather than on groups, whenever possible.

7. Application of Knowledge About Diet and Cancer to Programs in Public Health

Research to achieve this final strategic objective was proposed in an attempt to maximize the potential usefulness of public health programs to reduce the risk of cancer.

There is an urgent need to examine the behaviors and motivations of people who have already changed their diets, for example, in the direction suggested by "Dietary Guidelines for Americans"[15] or in this committee's first report.[10] Simultaneous study of those who have not changed their diets may facilitate an understanding of the obstacles to change. Although individuals who undertake "spontaneous" change may differ from the population as a whole, understanding their characteristics and identifying the periods in their lives when they changed their eating patterns would assist in understanding the factors that lead to long-term dietary change.[4]

Research is also needed to learn the "natural history" of dietary change in humans. For example, over the lifespans of most individuals, are there periods of vulnerability to change in food consumption when nutrition education might be most effective? Moore et al.[9] have recently suggested one potentially useful technique for such an investigation.

As stated earlier, better and more frequent monitoring of food intake and of patterns of food consumption in different groups is necessary to enhance our understanding of the effects of different diets and of long-term dietary change. Data are needed on a national basis about what people eat and how this varies in relation to such factors as geographical location, lifestyle, and ethnic and socio-economic factors. These data should be collected in such a manner that they can also be analyzed for information about the history of an individual's lifetime food habits, including changes in those habits.

Existing bodies of longitudinal dietary data, such as those from the Harvard[2] and Colorado[1] growth studies, should be examined for what they might reveal about patterns of food consumption over time and points at which diets change.

Consideration should be given to supporting additional long-term studies by researchers and/or observers to learn the factors affecting the acquisition of childhood eating patterns (see e.g., Thomas et al.[14]).

IV. COMMENT

This chapter summarizes useful directions proposed by the Committee on Diet, Nutrition, and Cancer for research on the relationship between diet, nutrition, and carcinogenesis. However, the committee was acutely cognizant of the widely held view among scientists that research is most vigorous when it is the product of individual decisions by investigators rather than the result of a preconceived strategy, and thus the above suggestions should be viewed in that context.

REFERENCES

1. **Beal, V. A.**, The nutritional history in longitudinal research, *J. Am. Diet. Assoc.*, 51, 426, 1967.
2. **Burke, B. S., Reed, R. B., van den Berg, A. S., and Stuart, H. C.**, Caloric and protein intakes of children between 1 and 18 years of age, *Pediatrics*, 24, 922, 1959.
3. Dietary Intake Source Data: United States, 1971—74, Publ. No. 79-1221, Public Health Service, National Center for Health Satistics, U. S. Department of Health and Human Services, Hyattsville, Md., 1979.
4. **Kasl, S. V.**, Cardiovascular risk reduction in a community setting: some comments, *J. Consult. Clin. Psychol.*, 48, 143, 1980.
5. **Keen, P. and Martin, P.**, The toxicity and fungal infestation of food stuffs in Swarziland in relation to harvesting and storage, *Trop. Geog. Med.*, 23, 35, 1971.
6. **King, C. G.**, Early experiences with ascrobic acid — a retrospect, *Nutr. Rev.*, 12, 1, 1954.
7. **Kraybill, H. F. and Shimkin, M. B.**, Carcinogenesis related to foods contaminated by processing and fungal metabolites, *Adv. Cancer Res.*, 8, 191, 1964.
8. **Logan, J. and Cairns, J.**, The secrets of cancer, *Nature (London)*, 300, 104, 1983.
9. **Moore, J. V., Prestridge, L. L., and Newell, G. R.**, Research techniques for epidemiologic investigation of nutrition and cancer, *Nutr. Cancer*, 3, 249, 1982.
10. National Research Council, Diet, Nutrition, and Cancer. Committee on Diet, Nutrition, and Cancer, Assembly of Life Sciences, National Research Council, National Academy Press, Washington, D. C., 1982.
11. **Palmer, S. and Bakshi, K.**, Diet, nutrition, and cancer: interim dietary guidelines, *J. Natl. Cancer Inst.*, 70, 1151, 1983.
12. National Research Council, Diet, Nutrition, and Cancer: Directions for Research, Commission on Life Sciences, National Academy of Sciences, Washington, D. C. 1983.
13. **Pott, P.**, Cancer scroti, in *Chirurgical Observations*, Howes, Clarke, and Collins, London, 1775, 63.
14. **Thomas, A., Chess, S., Birch, H. G., Hertzig, M. E., and Korn, S.**, *Behavioral Individuality in Early Childhood*, New York University Press, New York, 1963.
15. Nutrition and Your Health. Dietary Guidelines for Americans, Home and Garden Bulletin No. 232, U. S. Department of Agriculture and U. S. Department of Health, Education, and Welfare, Washington, D. C., 1985.
16. **White, A., Handler, P., Smith, E. L., Hill, R. L., and Lehman, R. I.**, The water soluble vitamins, *Principles of Biochemistry*, 6th Ed., McGraw-Hill, New York, 1978, 1333.
17. **Wogan, G.**, Aflatoxin carcinogenesis, in *Methods in Cancer Research*, Vol. 7, Busch, H., Ed., Academic Press, New York, 1973.

INDEX